T0140316

Artificial Intelligence: Foundations, Theory, and Algorithms

More information about this series at http://www.springer.com/series/13900

Christian Blum • Günther R. Raidl

Hybrid Metaheuristics

Powerful Tools for Optimization

Christian Blum
Dept. of Computer Science and
Artificial Intelligence
University of the Basque Country
San Sebastian, Spain

Günther R. Raidl
Algorithms and Data Structures Group
Vienna University of Technology
Vienna, Austria

ISSN 2365-3051 ISSN 2365-306X (electronic)
Artificial Intelligence: Foundations, Theory, and Algorithms
ISBN 978-3-319-80907-6 ISBN 978-3-319-30883-8 (eBook)
DOI 10.1007/978-3-319-30883-8

Softcover reprint of the hardcover 1st edition 2016

Printed on acid-free paper

This Springer imprint is published by Springer Nature
The registered company is Springer International Publishing AG Switzerland

To Gabi and María and our children Júlia, Manuela, Marc, and Tobias. Without their love and support life would not be the same.

Preface

Research in metaheuristics for combinatorial optimization problems has lately experienced a noteworthy shift towards the hybridization of metaheuristics with other techniques for optimization. At the same time, the focus of research has changed from being rather algorithm-oriented to being more problem-oriented. Nowadays the focus is on solving a problem at hand in the best way possible, rather than promoting a certain metaheuristic. This has led to an enormously fruitful cross-fertilization of different areas of optimization, algorithmics, mathematical modeling, operations research, statistics, simulation, and other fields. This cross-fertilization has resulted in a multitude of powerful hybrid algorithms that were obtained by combining components or concepts from different optimization techniques. Hereby, hybridization is not restricted to different variants of metaheuristics but includes, for example, the combination of mathematical programming, dynamic programming, constraint programming or statistical modeling with metaheuristics.

This book tries to cover several prominent hybridization techniques that have proven to be successful on a large variety of applications as well as some newer but highly promising strategies.

A first introductory chapter reviews basic principles of local search, prominent metaheuristics as well as tree search, dynamic programming, mixed integer linear programming, and constraint programming for combinatorial optimization purposes. The following chapters then present in detail five generally applicable hybridization strategies, including exemplary case studies on selected problems. These five approaches are:

- incomplete solution representations and decoders
- problem instance reduction
- large neighborhood search
- parallel non-independent construction of solutions within metaheuristics
- hybridization based on complete solution archives

While these strategies cover many or even most of the hybridization approaches used nowadays, there also exist several others. The last chapter therefore gives a brief overview on some further, prominent concepts and concludes the book.

San Sebastian – Donostia, Basque Country, Spain *Christian Blum*
Vienna, Austria *Günther R. Raidl*
September 2015

Acknowledgements

Christian Blum acknowledges support by grant TIN2012-37930-C02-02 of the Spanish Government. In addition, support is acknowledged from IKERBASQUE (Basque Foundation for Science). Experiments reported in this book were partly executed in the High Performance Computing environment managed by RDlab[1] and we would like to thank them for their support.

Günther Raidl's work is supported by the Austrian Science Fund (FWF) under grant P24660-N23.

Especially with respect to Chapters 2 and 6, we thank Bin Hu from the AIT Austrian Institute of Technology and Benjamin Biesinger from the Vienna University of Technology for their very valuable contributions. In particular, most of the reported experiments were conducted by them, and the text is based on joint publications with them.

Last but not least we would like to thank the restaurant of the *Sidreria Aginaga* for the inspiring effects of their *Sidreria Menu*.

[1] http://rdlab.lsi.upc.edu

Contents

Acronyms

ABC	Artificial Bee Colony
ACO	Ant Colony Optimization
AI	Artificial Intelligence
BD	Benders Decomposition
BS	Beam Search
CG	Column Generation
CO	Combinatorial Optimization
CP	Constraint Programming
DP	Dynamic Programming
EA	Evolutionary Algorithm
GA	Genetic Algorithm
GEEN	Global Edge Exchange Neighborhood
GESB	Global Edge Set Based
GLS	Guided Local Search
GMST	Generalized Minimum Spanning Tree
GRASP	Greedy Randomized Adaptive Search Procedures
IG	Iterated Greedy
IKH	Iterated Kruskal Based Heuristic (for the GMST problem)
ILP	Integer Linear Programming
ILS	Iterated Local Search
LAN	Local Area Network
LNS	Large Neighborhood Search
LD	Lagrangian Decomposition
LP	Linear Programming
MMAS	MAX-MIN Ant System
MCSP	Minimum Common String Partition
MDH	Minimum Distance Heuristic (for the GMST problem)
MDS	Minimum Dominating Set
MIP	Mixed Integer Programming
MKP	Multidimensional Knapsack Problem
MWDS	Minimum Weight Dominating Set

NEN Node Exchange Neighborhood
NSB Node Set Based
OR Operations Research
PSO Particle Swarm Optimization
R2NEN Restricted Two Node Exchange Neighborhood
RINS Relaxation Induced Neighborhood Search
SA Simulated Annealing
SMTWTSP Single-Machine Total Weighted Tardiness Scheduling Problem
SWO Squeaky Wheel Optimization
TS Tabu Search
UDG Unit Disk Graph
VND Variable Neighborhood Descent
VNS Variable Neighborhood Search

Chapter 1
Introduction

In the seminal work of Papadimitriou and Steiglitz [216] a combinatorial optimization (CO) problem $\mathscr{P} = (\mathscr{S}, f)$ is defined by means of a finite set of objects $\mathscr{S}$ and an objective function $f : \mathscr{S} \mapsto \mathbb{R}^+$ that assigns a non-negative cost value to each of the objects $s \in \mathscr{S}$. The process of optimization consists in finding an object s^* of minimal cost value. In this context, note that minimizing over an objective function f is the same as maximizing over $-f$, which is why every CO problem can be described as a minimization problem. The objects in $\mathscr{S}$ are typically integer vectors, subsets of a set of items, permutations of a set of items, or graph structures. A classic academic example is the well-known *Traveling Salesman Problem* (TSP). It has a completely connected, undirected graph $G = (V, E)$ with a positive weight assigned to each edge. The goal is to find, among all Hamiltonian cycles in G, one which is characterized by the lowest sum of the edge weights. Therefore, $\mathscr{S}$ consists of all possible Hamiltonian cycles in G, and the objective function value of a Hamiltonian cycle $s \in \mathscr{S}$ is defined as the sum of the weights of the edges appearing in s. While the TSP is famous in the academic literature, it admittedly does not that often appear in its pure form in practical transportation optimization applications. However, a huge number of related vehicle routing problems with additional aspects, like the need to consider more than one vehicle, vehicle capacities, time windows for visiting customers, shift times, etc., exist, and they are highly desired in practice. Unfortunately, most of these problems are very difficult to solve to optimality in practice, which is also documented in theoretical terms by the NP-hardness and non-approximability of the respective CO problems. Even slightly better solutions to such problems frequently imply a significant and direct financial, ecological and/or social impact and lead to substantial savings. Other prominent application areas where hard CO problems can be found are scheduling and timetabling, cutting and packing, network design (e.g., in telecommunication), location and assignment planning, and bioinformatics. The research area of combinatorial optimization is therefore a very important and active one, and methods yielding the best possible results are highly desired.

Due to the practical importance of CO problems, a large variety of algorithmic and mathematical approaches to tackle these problems has emerged in recent decades. These techniques can be fundamentally classified as being either *exact* or *approximate* methods. Exact algorithms guarantee to find for every finite size instance of a CO problem an optimal solution in bounded time [216, 203]. Yet—assuming that $P \neq NP$—an algorithm that solves a CO problem classified as being NP-hard in polynomial time does not exist [107]. As a consequence, such complete methods might need—in the worst-case—an exponential computation time for coming up with a proven optimal solution. This often leads to computation times too high for practical purposes. Thus, research on approximate methods to solve CO problems has enjoyed increasing attention in recent decades. Approximate methods sacrifice the guarantee of finding optimal solutions for the sake of being able to produce high-quality solutions in practically acceptable computation times.

1.1 Basic Heuristic Methods

In the following we first provide a short description of the most basic components of approximate algorithms: constructive heuristics and local search.

1.1.1 Constructive Heuristics

In the following we assume that, given a problem instance $\mathscr{I}$ to a generic CO problem $\mathscr{P}$, set C represents the set of all possible components of which solutions to the problem instance are composed. C is henceforth called the complete set of solution components with respect to $\mathscr{I}$. Moreover, a valid solution s to $\mathscr{I}$ is represented as a subset of the complete set C of solution components, i.e., $s \subseteq C$. Imagine, for example, the input graph G in the case of the TSP. The set of all edges can be regarded as the set C of all possible solution components. Moreover, the edges belonging to a tour s, i.e., a valid solution, form the set of solution components that are contained in s.

Constructive heuristics are generally said to be the fastest approximate methods. Solutions are generated (or constructed) from scratch by adding, at each step, a solution component to an initially empty partial solution. This is done until a solution is complete or another stopping criterion is satisfied. We henceforth assume that a solution construction stops when the current (partial) solution cannot be further extended. Constructive heuristics are characterized by a *construction mechanism*, which specifies for each partial solution s^{p} the set of possible extensions. This set, henceforth denoted by $Ext(s^{\mathrm{p}})$, is a subset of C, the complete set of solution components. At each construction step, one of the solution components from $Ext(s^{\mathrm{p}})$ is selected and added to s^{p}. This is done until $Ext(s^{\mathrm{p}}) = \emptyset$, which either means that s^{p} is a complete solution, or that s^{p} is a partial solution that cannot be extended to a complete solution. The algorithmic framework of a constructive heuristic is shown in Algorithm 1. Probably the most well-known example of a constructive heuristic

Algorithm 1 Constructive Heuristic

$s^{\mathrm{p}} = \emptyset$
while $Ext(s^{\mathrm{p}}) \neq \emptyset$ **do**
  $c \leftarrow$ Select($Ext(s^{\mathrm{p}})$)
  add c to s^{p}
end while

is a *greedy heuristic*, which makes use of a so-called greedy function in order to select a solution component from $Ext(s^{\mathrm{p}})$ at each step. A greedy function is a measure for the goodness of the solution components, i.e., an indicator for the impact on the objective value when a component would be included. Greedy heuristics choose at each step one of the extensions with the highest value—e.g., in case of the TSP possibly an edge with the lowest weight.

1.1.2 Local Search

In contrast to constructive heuristics that generate solutions from scratch, the main idea of *Local Search* (LS) is to start off from an initial solution and to replace the current solution, at each step, with a better solution from a suitably defined *neighborhood* of the current solution. This process stops as soon as the neighborhood of the current solution does not contain any solution which is better than the current one. Formally, the neighborhood of a solution s is derived by a so-called neighborhood function (also sometimes called neighborhood structure), which can be defined as follows.

Definition 1.1. A **neighborhood function** $\mathscr{N} : \mathscr{S} \rightarrow 2^{\mathscr{S}}$ assigns to each solution $s \in \mathscr{S}$ a set of neighbors $\mathscr{N}(s) \subseteq \mathscr{S}$, which is called the neighborhood of s.

Neighborhood functions are, in many cases, implicitly defined by specifying the changes that must be applied to a solution s in order to generate all its neighbors. The application of such an operator that produces a neighbor $s' \in \mathscr{N}(s)$ of a solution s is known as a *move*. Moreover, a solution $s^* \in \mathscr{S}$ is called a *global minimum* if for all $s \in \mathscr{S}$ it holds that $f(s^*) \leq f(s)$. The set of all global minima is henceforth denoted by $\mathscr{S}^*$. Given a neighborhood function $\mathscr{N}$, the concept of a local minimum can be defined as follows.

Definition 1.2. A **local minimum** with respect to a neighborhood function $\mathscr{N}$ is a solution $\hat{s}$ such that $\forall\, s \in \mathscr{N}(\hat{s}) : f(\hat{s}) \leq f(s)$. We call $\hat{s}$ a strict local minimum if $f(\hat{s}) < f(s)\ \forall\, s \in \mathscr{N}(\hat{s})$.

A basic local search method is sketched in Algorithm 2. There are two major ways of implementing the step function ChooseImprovingNeighbor($\mathscr{N}(s)$). The first way is known as *first improvement*. In *first improvement*, the neighborhood $\mathscr{N}(s)$ is examined in some pre-defined order and the first solution better than s is

Algorithm 2 Local Search

given: initial solution s, neighborhood function $\mathcal{N}$
while $\exists\, s' \in \mathcal{N}(s)$ such that $f(s') < f(s)$ **do**
  $s \leftarrow$ ChooseImprovingNeighbor($\mathcal{N}(s)$)
end while

returned. In contrast, *best improvement* refers to an exhaustive exploration of the neighborhood which returns a best solution found in $\mathcal{N}(s)$. Note that with both methods a local search algorithm stops at a local minimum.

In general, the performance of a local search method strongly depends on the definition of the neighborhood function $\mathcal{N}$. Finally, note that a local search algorithm partitions the search space $\mathcal{S}$ into so-called *basins of attraction* of local minima. Hereby, the basin of attraction of a local minimum $\hat{s} \in \mathcal{S}$ is the set of all solutions s for which the corresponding local search method terminates in $\hat{s}$ when started from s as initial solution. For what concerns the relation to constructive heuristics, we can state that constructive heuristics are often faster than local search methods, yet they frequently return solutions of inferior quality.

1.1.2.1 Variable Neighborhood Descent

A natural way of extending basic deterministic local search is to consider not just a single neighborhood function but a finite set of multiple neighborhood functions $\{\mathcal{N}_1, \ldots, \mathcal{N}_{k_{\max}}\}$. *Variable Neighborhood Descent* (VND) [127] does so and systematically explores them until a solution is obtained that is locally optimal w.r.t. each of the considered neighborhood functions. In principle, it is not required that there is a relation between the neighborhood functions. Frequently, however, the neighborhood functions are ordered in some way so that for all $s \in \mathcal{S}$ it holds that $|\mathcal{N}_1(s)| \leq |\mathcal{N}_2(s)| \leq \ldots \leq |\mathcal{N}_{k_{\max}}(s)|$. Note that this does not imply that $\mathcal{N}_i(s) \subset \mathcal{N}_{i+1}(s)$, for $i = 1, \ldots, k_{\max} - 1$, which should be avoided.

The algorithm works as follows. Given a starting solution $s \in \mathcal{S}$, it chooses at each iteration an improving or a best neighbor s' from a neighborhood $\mathcal{N}_k(s)$, $k = 1, \ldots, k_{\max}$. Hereby, index k is initialized with one at the start of the algorithm. If s' is better than s, s is accepted as new incumbent solution and k is reset to one. Otherwise, k is incremented. The algorithm terminates when $k > k_{\max}$, i.e., when no better solution could be found in any of the $k_{\max}$ neighborhoods of s.

1.2 Metaheuristics

In the '70s, a new class of general approximate algorithms, called *metaheuristics*, emerged that basically try to combine constructive heuristics and/or local search methods with other ideas in higher-level frameworks aimed at effectively explor-

Algorithm 3 Variable Neighborhood Descent (VND)

given: initial solution s, neighborhood functions $\mathcal{N}_k$, $k = 1,\ldots,k_{\max}$
$k \leftarrow 1$
while $k \leq k_{\max}$ **do**
  $s' \leftarrow$ ChooseBestNeighbor($\mathcal{N}_k(s)$)
  if $f(s') < f(s)$ **then**
    $s \leftarrow s'$
    $k \leftarrow 1$
  else
    $k \leftarrow k+1$
  end if
end while

ing a search space in order to find an optimal or near-optimal solution. The origins of metaheuristics are to be found in the Artificial Intelligence and Operations Research communities [250, 113, 31]. In general, they are approximate algorithms for optimization that are not specifically expressed for a particular problem. Ant colony optimization, variable neighborhood search, genetic and evolutionary algorithms, iterated local search, simulated annealing and tabu search (in alphabetical order) are prominent representatives of the class of metaheuristic algorithms. Each of these metaheuristics has its own historical background. While some metaheuristics are inspired by natural processes such as evolution or the shortest-path-finding behavior of ant colonies, others are extensions of less sophisticated algorithms such as greedy heuristics and local search [137]. An important point of metaheuristics is that they all include mechanisms for escaping from local minima. In other words, they are extensions of constructive heuristics and local search methods with the aim of *exploring* the search space of the tackled problem instance in less limited ways.

In the following we provide short descriptions of the basic ideas underlying a selection of the most prominent metaheuristics known to date. Many different variants, however, have also emerged from each of them in the meantime.

1.2.1 Greedy Randomized Adaptive Search Procedures

The *Greedy Randomized Adaptive Search Procedure* (GRASP) [92, 230] is one of the simplest metaheuristics. It combines a randomized greedy construction of solutions with the subsequent application of local search; see Algorithm 4. The best found solution is returned when the algorithm stops due to the termination conditions being met. The solution construction in GRASP frequently works as follows. Given a partial solution s^{p} and the set $Ext(s^{\text{p}})$ of solution components that may be used to extend s^{p}, a restricted candidate list $L \subseteq Ext(s^{\text{p}})$ is determined that contains the best solution components—with respect to a greedy function—of $Ext(s^{\text{p}})$. Then, a solution component $c^* \in L$ is chosen at random. The length α of the restricted candidate list L determines the strength of the bias that is introduced by the

Algorithm 4 Greedy Randomized Adaptive Search Procedure (GRASP)

```
while termination conditions not met do
    s ← ConstructGreedyRandomizedSolution()
    s ← LocalSearch(s)
end while
```

greedy function. In the extreme case of $\alpha = 1$ a best solution component would always be added deterministically at each construction step, which would result in the construction of the same solution as done by the corresponding greedy heuristic. On the other side, choosing $\alpha = |Ext(s^{\mathrm{p}})|$ results in a random solution construction without any heuristic bias. Thus, α is a critical parameter in GRASP that controls the diversity of the solution creation. It might be set to a fixed integer or determined by an adaptive scheme, depending on each solution construction step.

The second phase of the algorithm consists in the application of a local search method to the constructed solution. The options for this local search method range from a basic local search or variable neighborhood descent to other metaheuristics such as iterated local search (see Section 1.2.3) or tabu search (see Section 1.2.5). For GRASP to be effective, at least two conditions should be satisfied: (1) the solution construction mechanism should sample promising regions of the search space, and (2) the constructed solutions should be good starting points for local search.

1.2.2 Iterated Greedy Algorithms

Iterated Greedy (IG) algorithms [262] are metaheuristics based on another relatively simple idea: At each iteration, the incumbent solution s is partially destroyed, resulting in a partial solution s^{p}. This destruction process is usually done in a probabilistic way, potentially guided by some measure about the usefulness of solution components–that is, solution components for which this measure is rather low have a higher chance to be removed from the incumbent solution s. After this partial destruction the resulting partial solution s^{p} is subject to a re-generation process which produces a new complete solution s' that contains s^{p}. For this step, usually, the same greedy construction mechanism is used as for the generation of the initial solution. Afterwards, an optional local search method may be applied to s'. However, most IG algorithms published in the literature do not make use of this feature, as one of the key strengths of these algorithms is the high computational efficiency resulting from the fact of being relatively unsophisticated from an algorithmic point of view. Finally, the resulting solution $\hat{s'}$ may either be accepted as new current solution s, or not. This is decided in function ApplyAcceptanceCriterion($\hat{s'}, \hat{s}, history$). Two extreme examples are (1) accepting the new solution only in case of an improvement and (2) always accepting the new solution. In between, there are several possibilities. For example, an acceptance criterion that is similar to that of simulated annealing, which will be described in Section 1.2.4, can be adopted.

Algorithm 5 Iterated Greedy (IG) Algorithm

1: $s \leftarrow$ GenerateInitialSolution()
2: **while** termination conditions not met **do**
3: $s^{p} \leftarrow$ DestroyPartially(s)
4: $s' \leftarrow$ RecreateCompleteSolution(s^{p})
5: $\hat{s}' \leftarrow$ LocalSearch(s') {optional}
6: $s \leftarrow$ ApplyAcceptanceCriterion($s, \hat{s}'$)
7: **end while**

According to Hoos and Stützle [137], two important advantages of starting the solution construction from partial solutions are that (1) the solution construction process is much faster and (2) good parts of solutions may be exploited directly. The pseudo-code of IG is shown in Algorithm 5.

1.2.3 Iterated Local Search

Iterated Local Search (ILS) [275, 181] is another metaheuristic based on a simple idea. Instead of repeatedly applying local search to independently generated starting solutions as in GRASP, an ILS algorithm produces the starting solution for the next iteration by randomly *perturbing* an incumbent solution. Perturbation should move the search to a different basin of attraction, but at the same time cause the search to stay close to the current incumbent solution.

The pseudo-code of ILS is shown in Algorithm 6; it works as follows. First, an initial solution is generated in function GenerateInitialSolution(). This solution is subsequently improved by the application of local search in function LocalSearch(s). The construction of initial solutions should be fast (i.e., computationally inexpensive), and—if possible—initial solutions should be a good starting point for local search. At each iteration, the incumbent solution $\hat{s}$ is perturbed in function Perturbation($\hat{s}$,*history*), possibly depending on the search history. This results in a perturbed solution s'. The perturbation is usually non-deterministic in order to avoid cycling, and the importance of choosing its strength carefully is obvious: On the one side, a perturbation that makes too few changes might not enable the algorithm to escape from the basin of attraction of the current solution. On the other side, a perturbation that is too strong would make the algorithm similar to a random restart local search. Finally, an acceptance criterion like in the case of IG (see the previous Section 1.2.2) is employed for choosing the incumbent solution for the next iteration from s and $\hat{s}'$. In more sophisticated variants of ILS, the acceptance criterion might depend—like the perturbation mechanism—on the history of the search process.

Algorithm 6 Iterated Local Search (ILS)

```
1: s ← GenerateInitialSolution()
2: ŝ ← LocalSearch(s)
3: while termination conditions not met do
4:     s′ ← Perturbation(ŝ, history)
5:     ŝ′ ← LocalSearch(s′)
6:     ŝ ← ApplyAcceptanceCriterion(ŝ′, ŝ, history)
7: end while
```

1.2.4 Simulated Annealing

Simulated Annealing (SA) is one of the oldest (if not the oldest) metaheuristic. In any case, SA is surely one of the first algorithms with an explicit strategy for escaping from local minima. The algorithm has its origins in statistical mechanics and the Metropolis algorithm [194]. The main idea of SA is based on the annealing process of metal and glass. After bringing both materials to the fluid state, the crystal structure of both materials assumes a low energy configuration when cooled with an appropriate cooling schedule. SA was first presented as a search algorithm for CO problems in [157] and [52]. In order to escape from local minima, the fundamental idea is to allow moves to solutions with objective function values that are worse than the objective function value of the current solution. Such a kind of move may also be called an *uphill move*. At each iteration a solution $s' \in \mathcal{N}(s)$ is randomly chosen. If s' is better than s (i.e., has a lower objective function value), then s' is accepted as the new current solution. Otherwise, if the move from s to s' is an uphill move, s' is accepted with a probability which is a function of a temperature parameter T_k and $f(s') - f(s)$. Usually this probability is computed following the Boltzmann distribution. Moreover, during a run of SA, the value of T_k generally decreases. In this way, the probability of accepting a solution that is worse than the current one decreases during a run.

It is interesting to note that the dynamic process performed by SA can be modeled as a *Markov chain* [91], as it follows a trajectory in the state space in which the successor state is chosen depending only on the incumbent one. This means that basic SA is a memory-less process. The algorithmic framework of SA is sketched in Algorithm 7.

1.2.5 Tabu Search

Like SA, *Tabu Search* (TS) [114] is one of the older metaheuristics. The basic idea of TS is the explicit use of search history, both to escape from local minima and to have a mechanism for the exploration of the search space.

A basic TS as shown in Algorithm 8 is based on *best-improvement* local search (see Section 1.1.2). The main idea is to make use of *short-term memory* to escape

Algorithm 7 Simulated Annealing (SA)

$s \leftarrow$ GenerateInitialSolution()
$k \leftarrow 0$
$T_k \leftarrow$ SetInitialTemperature()
while termination conditions not met **do**
  $s' \leftarrow$ SelectNeighborAtRandom($\mathcal{N}(s)$)
  if $(f(s') < f(s))$ **then**
    $s \leftarrow s'$
  else
    accept s' as new solution with a probability $\mathbf{p}(s' \mid T_k, s)$
  end if
  $T_{k+1} \leftarrow$AdaptTemperature(T_k, k)
  $k \leftarrow k+1$
end while

from local minima and to avoid sequences of moves that constantly repeat themselves, that is, *cycling*. In our basic TS, the short-term memory is implemented in terms of a *tabu list TL* of limited length in order to keep track of the most recently visited solutions. More specifically, the tabu list is used to exclude them from the neighborhood of the current solution. The resulting restricted neighborhood of a solution s will henceforth be denoted by $\mathcal{N}_a(s)$. At each iteration the best solution from the restricted neighborhood is chosen as the new current solution. Furthermore, procedure Update(TL,s,s') appends this solution to the tabu list. When *TL* has reached its maximum capacity, its oldest solution is removed. This allows the TS to possibly get back later to this solution in order to move from it to other neighboring solutions that are then not blacklisted in *TL*. With an infinite *TL*, the algorithm would run into the danger of blocking paths to certain regions of the search space forever, making them effectively unreachable. The TS stops when a termination condition is met.

The length l of the tabu list—also known as the *tabu tenure*—controls the memory of the search process. Rather small tabu tenures have the effect of focusing the search on rather confined areas of the search space. In contrast, rather large tabu tenures force the search process to do a more diversified search because a higher number of solutions is forbidden to be be revisited.

It has been observed by several authors that varying the tabu tenure during the search process generally leads to more robust algorithms. An example can be found in [279]. In this work, the tabu tenure is periodically reinitialized at random from an interval $[l_{\min}, l_{\max}]$. A more sophisticated way of managing a dynamic tabu tenure is based on the search history. In the presence of evidence for repeated solutions, for example, the tabu tenure may be increased [13, 12], and vice versa. Such TS variants are frequently called *Reactive TS*. Other advanced mechanisms of dynamically controlling tabu tenures are described in [111].

Finally, we have to remark that a crucial weakness of the above described basic TS is the overhead introduced for storing complete solutions in the tabu list and checking each encountered neighboring solution against the whole tabu list. This overhead may be a substantial burden in terms of memory, but even more importantly, in terms of computation time for comparing solutions, making the whole TS

Algorithm 8 Tabu Search (TS)

$s \leftarrow$ GenerateInitialSolution()
$TL \leftarrow \emptyset$
while termination conditions not met **do**
  $\mathcal{N}_a(s) \leftarrow \mathcal{N}(s) \setminus TL$
  $s' \leftarrow \text{argmin}\{f(s'') \mid s'' \in \mathcal{N}_a(s)\}$
  Update(TL,s,s')
  $s \leftarrow s'$ {s' always replaces s}
end while

ineffective. This even holds when implementing the tabu list relatively efficiently–e.g., via a hash table in order to avoid $O(l)$ comparisons for each check of whether or not a solution is tabu.

There is, however, a typically simple option to make the whole approach much more efficient, which is usually realized in practical TS implementations: Instead of complete solutions, only information of the performed moves to reach the encountered solutions is stored in the tabu list, and reversing the corresponding moves is forbidden. It is clear, however, that forbidding a certain move will in general disallow a larger set of solutions, and this must be considered when choosing the tabu tenure.

1.2.6 Variable Neighborhood Search

Variable Neighborhood Search (VNS) was first proposed in [126, 127]. It can be seen as a probabilistic variant or extension of the deterministic VND method outlined in Section 1.1.2. The essential feature of VNS is the fact that—just like VND—it explicitly applies strategies for swapping between different neighborhood functions. In contrast to VND, however, it does so for diversifying the search and escaping from local optima in a controlled manner.

The algorithmic framework of VNS is sketched in Algorithm 9. As input, the algorithm requires a finite set of neighborhood functions for the tackled problem. As already mentioned in the context of VND, it is not required that there is a relation between these neighborhood functions. However, often they are ordered according to increasing neighborhood size. In comparison to VND, the respective neighborhoods are typically larger as they are not searched in a first or best-improvement manner but only randomly sampled, as we will see. Frequently, these neighborhood functions are therefore defined by *compound moves*, i.e., a certain number of repetitions of a basic move type.

The algorithm works as follows. After the generation of an initial solution, each iteration consists of three phases: *shaking*, *local search* and *move*. The shaking phase serves for randomly selecting a solution s' from the k-th neighborhood of the current solution s. Next, a local search algorithm is applied to s'. This local search algorithm may use any neighborhood function, independently from $\mathcal{N}_1, \ldots, \mathcal{N}_{k_{\max}}$.

Algorithm 9 Variable Neighborhood Search (VNS)

select a set of neighborhood functions $\mathcal{N}_k, k = 1, \ldots, k_{\max}$
$s \leftarrow$ GenerateInitialSolution()
while termination conditions not met **do**
  $k \leftarrow 1$
  while $k < k_{\max}$ **do**
    $s' \leftarrow$ PickAtRandom($\mathcal{N}_k(s)$) {also called *shaking phase*}
    $s'' \leftarrow$ LocalSearch(s')
    if $f(s'') < f(s)$ **then**
      $s \leftarrow s''$
      $k \leftarrow 1$
    else
      $k \leftarrow k + 1$
    end if
  end while
end while

The resulting local minimum s'' is compared to s, the incumbent solution, and if it is of higher quality it replaces s. Moreover, in this case the algorithm proceeds with the first neighborhood function, i.e., k is set to one. Otherwise, k is incremented for the shaking phase of the next iteration. The goal of the shaking phase is to move the search process to a different local minimum with respect to the employed local search algorithm. However, the solution selected by shaking should not be "too different" from s, because otherwise the algorithm would resemble multistart local search. Choosing s' randomly from a suitable neighborhood of the current solution is assumed to be likely to produce a solution that maintains some features of the current one.

We can see here a close relationship of shaking to the perturbation operation in ILS. However, VNS implements this operation in a more sophisticated way. If the current incumbent solution is not improved, the neighborhood for shaking is changed with the aim of increasing the diversification. This is particularly the case when the corresponding neighborhoods are of increasing cardinality.

The effectiveness of the dynamic strategy employed by VNS (and VND) for swapping between neighborhood functions can be explained by the fact that a "bad" place on the search landscape with respect to one neighborhood function might be a "good" place on the search landscape with respect to a different neighborhood function. Moreover, a local minimum with respect to a neighborhood function is possibly not a local minimum with respect to another neighborhood function.

Occasionally, a VND with its own set of neighborhood functions also is applied as advanced local search within a VNS with a different set of shaking neighborhood functions. Such an approach is called *Generalized VNS* [127].

Algorithm 10 Ant Colony Optimization (ACO)

```
1: while termination conditions not met do
2:    ScheduleActivities
3:       AntBasedSolutionConstruction()
4:       PheromoneUpdate()
5:       DaemonActions() {optional}
6:    end ScheduleActivities
7: end while
```

1.2.7 Ant Colony Optimization

The development of the *Ant Colony Optimization* (ACO) metaheuristic [78] was inspired by the observation of the shortest-path-finding behavior of natural ant colonies. From a technical perspective, ACO algorithms work as follows. Given the complete set of solution components C with respect to the tackled problem instance, a set $\mathcal{T}$ of *pheromone values* must be defined.[1] This set is commonly called the *pheromone model*, which is—from a mathematical point of view—a parameterized probabilistic model. The pheromone model is one of the central components of any ACO algorithm. Pheromone values $\tau_i \in \mathcal{T}$ are commonly associated with solution components. The pheromone model is used to probabilistically generate solutions to the problem under consideration by assembling them from the set of solution components. In general, ACO algorithms attempt to solve an optimization problem by iterating the following two steps:

- candidate solutions are constructed using a pheromone model, that is, a parameterized probability distribution over the search space;
- the candidate solutions are used to update the pheromone values in a way that is deemed to bias future sampling toward high-quality solutions.

The pheromone update aims to concentrate the search in regions of the search space containing high-quality solutions. In particular, the reinforcement of solution components depending on the solution quality is an important ingredient of ACO algorithms. It implicitly assumes that good solutions consist of good solution components. Learning which components contribute to good solutions can help assemble them into better solutions. The main steps of any ACO algorithm are shown in Algorithm 10. Daemon actions (see line 5 of Algorithm 10) may include, for example, the application of local search to solutions constructed in function AntBasedSolutionConstruction().

The class of ACO algorithms comprises several variants. Among the most popular ones are MAX-MIN Ant System (MMAS) [277] and Ant Colony System (ACS) [77]. For more comprehensive information we refer the interested reader to [79].

[1] See also the first paragraph of Section 1.1.1 for the definition of C.

Algorithm 11 Evolutionary Algorithm (EA)

1: $P \leftarrow$ GenerateInitialPopulation()
2: **while** termination conditions not met **do**
3: $P^{s} \leftarrow$ Selection(P)
4: $P^{off} \leftarrow$ Recombination(P^{s})
5: $P' \leftarrow$ Mutation(P^{off})
6: $P \leftarrow$ Replacement(P, P')
7: **end while**

1.2.8 Evolutionary Algorithms

Algorithms that are based on principles of natural evolution are called *Evolutionary Algorithms* (EAs) [11]. EAs can be characterized as strongly simplified computational models of evolutionary processes. They are inspired by nature's capability to evolve living beings well adapted to their environment. At the core of each EA is a *population P* (set) of *individuals*, which are generally candidate solutions to the considered optimization problem. After generating an initial population of diverse individuals, which is usually done in a randomized way, at each iteration a subset P^{s} of individuals is selected from the current population P. These individuals serve as *parents* for one or more *reproduction* operators in order to generate a set P^{off} of offspring individuals. EAs typically use randomized reproduction operators called *recombination* or *crossover* to derive new individuals from the properties (e.g., solution components) contained in the selected parents. EAs usually also make use of *mutation* or *modification* operators that cause smaller "self-adaptive" modifications of the individuals in P^{off}, resulting in a set P' of individuals. This self-adaption frequently has a completely random character, but is occasionally also heuristically guided. Finally, the population for the next iteration is selected from the old population P and the offsprings P'. This process is sketched in Algorithm 11.

The driving force in EAs is the *selection* of individuals based on their *fitness*, which is a measure that is generally based on the objective function, the result of simulation experiments, or some other kind of quality measure. Individuals with higher fitness typically have a higher probability to be chosen as members of the population of the next generation (or as parents for the generation of new individuals). This corresponds to the principle of *survival of the fittest* in natural evolution. It is the capability of nature to adapt itself to a changing environment, which gave the inspiration for EAs.

There has been a variety of different EAs proposed over the decades. Three different strands of EAs developed independently in the early years. These are *Evolutionary Programming* (EP) as introduced by Fogel in [102] and Fogel et al. in [103], *Evolutionary Strategies* (ESs) proposed by Rechenberg in [249] and *Genetic Algorithms* (GAs) initiated by Holland in [133]; see [116], [197], [251], and [290] for further references. EP arose from the desire to generate machine intelligence. While EP originally was proposed to operate on discrete representations of finite state machines, most of the present variants are used for continuous optimization problems.

The latter also holds for most present variants of ESs, whereas GAs are mainly applied to solve discrete problems. Later, other members of the EA family, such as *Genetic Programming* (GP) [165] and *Scatter Search* (SS) [115], were developed. Despite this division into different strands, EAs can be basically understood from a unified point of view with respect to their main components and the way they explore the search space. Over the years there have been quite a few overviews and surveys about EAs, including those by Bäck [10], by Fogel et al. [101], by Kobler and Hertz [132], by Spears et al. [274], and by Michalewicz and Michalewicz [195]. In [49] a taxonomy of EAs is proposed.

Frequently, EAs are also hybridized with LS by trying to locally improve some or all of the generated offspring. As EAs are commonly known to have good exploration capabilities but are weak in fine-tuning solutions, such an approach often makes perfect sense. Such LS-based EAs are also referred to as *memetic algorithms* [201].

1.2.9 Further Metaheuristics

Among other well-known metaheuristics which were not sketched in previous subsections is, for example, *Particle Swarm Optimization* (PSO) [156, 60], which is a swarm intelligence technique for optimization inspired by the collective behavior of flocks of birds and/or schools of fish. The first PSO algorithm was introduced in 1995 by Kennedy and Eberhart [155] for the purpose of optimizing the weights of a neural network, thus, for continuous optimization, but has in the meantime also been adapted for approaching discrete problems (see, for example, [214]).

The *Artificial Bee Colony* (ABC) metaheuristic was first proposed in [151, 152]. The inspiration for the ABC algorithm is to be found in the foraging behavior of honey bees, which consists essentially of three components: food source positions, amount of nectar and three types of honey bees, that is, *employed bees*, *onlookers* and *scouts*. Essentially, the difference between the ABC algorithm and other population-based optimization techniques is to be found in the specific way of managing the resources of the algorithm, as suggested by the foraging behavior of honey bees. As in the case of PSO, ABC was initially introduced for continuous optimization. However, in the meantime the algorithm has been adapted for the application to combinatorial optimization problems as well (see, for example, [215, 256]).

One of the older metaheuristics, which has not received so much attention in recent years, is *Guided Local Search* (GLS) [292, 291]. We mention this metaheuristic because it applies a strategy for escaping from local minima that is quite different from those of other local search-based metaheuristics: GLS dynamically changes the objective function by additional terms based on the search history, resulting in a dynamically modified search landscape. The aim is to make already reached local minima gradually *less desirable* over time.

Apart from the metaheuristics for combinatorial optimization that were mentioned so far, many other ones have been applied rather sporadically in the literature.

Examples are *Squeaky Wheel Optimization* [147], *Extremal Optimization* [38], and the *Great Deluge Algorithm* [82].

After the above introduction to (meta)heuristic methods, we get back to exact techniques in the following sections. We will briefly overview some of the most prominent main principles, provide references for further reading, and in particular focus on those techniques that will be used in our hybrid metaheuristics in the further chapters of this book. As already mentioned, the applicability of exact techniques is in practice unfortunately often limited to small or medium-sized instances due to the complexity of the problems at hand. Nevertheless, one should not underestimate them. In particular, the property that a given CO problem is NP-hard does not immediately imply that an exact method cannot be used in practice. Instances relevant in practice may either be small enough or have certain structures that might be exploited in order to obtain proven optimal solutions effectively in a short time. For example, even huge instances of the NP-hard knapsack problems can frequently be efficiently solved to proven optimality by dynamic programming, as we will see in Section 1.3. Beware that NP-hardness just implies that there exists at least one instance for which we do not know how to solve it in polynomial time, but many or even most instances might be efficiently solvable. We therefore recommend in a practical setting to start by first considering traditional exact possibilities to solve a problem at hand, before turning to heuristics and hybrid approaches.

1.3 Tree Search Methods

The probably most basic approach for finding an optimal solution to a CO problem is to enumerate all potential solutions in order to identify one with the best objective value. Clearly such a naive *exhaustive enumeration* is typically inefficient and impractical, but it provides a starting point. Such an enumeration of all potential solutions is most typically done by following the principle of *divide and conquer* realized in the form of a *tree search*.

In tree search algorithms, the search space is in general recursively partitioned into two or more usually disjunct subspaces. Each partitioning, which is also called *branching*, is in general achieved by adding appropriate constraints for the respective subspaces. Most frequently, the domain of a selected variable is restricted for this purpose. For example, a binary variable might be either fixed to 0 or 1, or more generally, the domain of a variable $x \in \{a,\dots,b\}$ might be split by specifying $x \in \{a,\dots,t\}$ for one subspace and $x \in \{t+1,\dots,b\}$ for the second subspace, with $t \in \{a+1,\dots,b-1\}$. An exhaustive naive tree search performs this recursive partitioning until single solutions are obtained (or empty subproblems, if they cannot be easily avoided).

The name tree search comes from the following observation: We can associate the recursive partitioning with a rooted directed tree in which the root represents the whole search space and each subnode a respective subspace (subproblem). This tree

Algorithm 12 Branch-and-Bound

$\Pi \leftarrow \{\mathrm{P}\}$ {open subproblem list initialized with original problem}
$U \leftarrow \infty$ {global upper bound}
while $\Pi \neq \emptyset$ **do**
  select subproblem $P' \in \Pi$; $\Pi \leftarrow \Pi \setminus \{P'\}$
  {bounding:}
  $L_{P'} \leftarrow$ determine lower bound for P'
  if $L_{P'} < U$ **then**
    $x \leftarrow$ try to determine heuristic solution $x \in P'$
    if $\exists x \wedge f(x) < U$ **then**
      $x^* \leftarrow x$; $U \leftarrow f(x)$ {new best solution}
    end if
  end if
  if $L_{P'} < U$ **then**
    {branching:}
    partition P' into disjoint nonempty subproblems $P_1, \ldots, P_k$
    $\Pi \leftarrow \Pi \cup \{P_1, \ldots, P_k\}$
  end if
end while
return x^*

is called the *search tree*, and we can apply different graph traversal strategies on it for searching for a best solution. Basic strategies are *depth-first*, where a successor of a node—if one exists—is immediately considered before tracking back to the last predecessor with further, still unconsidered nodes, and *breadth-first*, where all nodes at a certain level are processed before considering the nodes at the next deeper level.

Obviously, exhaustive enumeration methods that consider all nodes of a search tree up to those corresponding to individual solutions are in general prone to excessive running times for larger search spaces. Advanced tree search techniques therefore apply a variety of techniques for *pruning* the search, i.e., recognizing subproblems (nodes) in which an optimal solution cannot appear as early as possible and skipping its processing. Clearly, pruning nodes closer to the root will in general imply much stronger time-savings than pruning nodes at deeper levels.

Algorithm 12 illustrates this basic principle of branch-and-bound for solving a minimization problem P. Π represents here the list of open subproblems, U the global upper bound, P' a selected subproblem to be processed, $L_{P'}$ the lower bound obtained for subproblem P', x a possible heuristic solution, $f(x)$ its objective value and x^* the current incumbent and finally optimal solution.

In general, there are several major design decisions for realizing an effective branch-and-bound. Besides the methods for determining lower and upper bounds, also the way of branching—e.g., which variable to select for partitioning the search space and how to split the domain—and the selection of the next subproblem to be processed play particularly important roles.

For further information on tree search and branch-and-bound in general, we recommend the textbooks [159, 65, 271].

Algorithm 13 Fibonacci (n)

```
if n = 1 ∨ n = 2 then
    return 1
else
    return Fibonacci(n − 1) + Fibonacci(n − 2)
end if
```

Algorithm 14 Linear Time Fibonacci (n)

```
F[1] ← 1
F[2] ← 1
for i ← 3,...,n do
    F[i] ← F[i−1] + F[i−2]
end for
return F[n]
```

1.4 Dynamic Programming

Similarly to branch-and-bound and other tree search approaches, *Dynamic Programming* (DP) also follows the principle of divide and conquer, i.e., a complex problem is recursively broken down into simpler subproblems whose solutions are then used to come up with a solution for the whole problem. In contrast to these other methods, however, DP can be particularly efficient on problems exhibiting *overlapping subproblems*, as one of its main principles is to store results to subproblems in order to possibly reuse them, avoiding repeated solving of the same subproblems. This principle is called *memoization*. For deeper introductions to DP we recommend the respective chapters in the textbooks [159, 65], from where the following examples regarding DP are also adopted.

Let us start with a simple example. The sequence of *Fibonacci numbers* is defined as follows:

$$F_1 = F_2 = 1, \qquad F_n = F_{n-1} + F_{n-2} \quad \forall n >= 3. \tag{1.1}$$

Clearly, this definition can be directly translated into a recursive procedure for calculating the i-th Fibonacci number as shown in Algorithm 13.

This realization, however, is not particularly efficient as the total number of calls of algorithm Fibonacci for determining the i-th Fibonacci number corresponds to the i-th Fibonacci number itself and the algorithm's overall time complexity is exponential. A much better approach is to store determined smaller Fibonacci numbers and to reuse them, yielding the linear time Algorithm 14.

Applying this principle to combinatorial optimization problems exhibiting a suitable structure with overlapping subproblems may lead to very efficient solving methods. Prominent examples are shortest path algorithms on graphs, calculating certain edit distances on strings, and knapsack problems. We use the latter to illustrate DP in more detail.

Algorithm 15 Dynamic Programming for Knapsack Problem (v,w,W)

```
v[0] ← 0
x[0] ← (0,...,0)
for j ← 1,...,W do
  v[j] ← v[j−1]
  x[j] ← x[j−1]
  for i ← 1,...,n do
    if w_i ≤ j ∧ v[j−w_i] + v_i > v[j] then
      v[j] ← v[j−w_i] + v_i
      x[j]_i ← x[j]_i + 1
    end if
  end for
end for
return v[W], x[W]
```

Knapsack Problem (with Repetition)

Given are n types of items $i = 1,\ldots,n$ with positive integer weights $w_i \in \mathbb{N}$ and values $v_i \in \mathbb{N}$ and a knapsack that can carry arbitrary many items with a total weight of at most W. Let us denote the number of items packed from each type by a solution vector $x = (x_1,\ldots,x_n) \in \mathbb{N}_0^n$. The objective is to maximize the value of packed items

$$v(x) = \sum_{i=1}^{n} v_i x_i \tag{1.2}$$

under the capacity constraint (*knapsack constraint*)

$$\sum_{i=1}^{n} w_i x_i \leq W. \tag{1.3}$$

We apply DP to the knapsack problem by considering as subproblems instances with smaller knapsack capacities. More specifically, let $v^*(W)$ be the optimal solution value for the knapsack problem instance with capacity W. It can be recursively determined by

$$v^*(W) = \max_{i=1,\ldots,n \mid w_i \leq W} \{v^*(W - w_i) + v_i\}, \tag{1.4}$$

with the assumption that the maximum over an empty set is zero. This recursion translates into the following algorithm using memoization to avoid recalculations:

In this algorithm array $v[0],\ldots,v[W]$ finally contains the optimal solution values for all capacities 0 to W and $x[0],\ldots,x[W]$ the corresponding solution vectors indicating the packed items.

The total runtime of the algorithm is in $O(Wn)$, that is, in principle linear in n. This holds despite the fact that the knapsack problem is NP-hard, which might appear surprising at first glance. Note, however, that the constant W depends on the instance and may be very large, somehow hiding the fact that the actual runtime may be long. Such algorithms are called *pseudo-polynomial* and do not contradict the fundamental assumption that NP-hard problems can in general not be solved in polynomial time. Here, knapsack instances may be obviously constructed in which W is chosen in an exponential relationship to the number of items n. Nevertheless, in practice DP approaches based on the sketched principle can be extremely efficient also for large knapsack problems, especially when the weights are generally small or, more precisely, when the number of discrete weight values that can be reached by feasible subsets of items is small.

To conclude, DP is an important class of algorithms that may work very well for problems exhibiting a certain structure, i.e., can be expressed in recursive ways with overlapping subproblems. Unfortunately, many practical combinatorial optimization problems remain that do not have such a structure, or where at least effective DP approaches are not known. As we will see for example in Chapter 2, hybrid metaheuristics may occasionally nevertheless provide possibilities of utilizing DP in meaningful ways.

1.5 Mixed Integer Linear Programming

Mixed Integer (Linear) Programming (MIP) is in practice very popular to approach combinatorial optimization problems as it is frequently not that difficult to come up with a valid mathematical model for a given problem that consists of a linear objective function and a set of linear constraints on continuous and/or integral variables. The practical success lies mainly in the powerful generic MIP solvers that exist today, such as CPLEX Optimizer[2], GUROBI[3], SCIP[4], or XPRESS[5]. Having a suitable model, these solvers are frequently able to solve instances of difficult problems in reasonable time. In contrast to (meta)heuristics, MIP-solvers are complete optimization techniques in the sense that they are able to return proven optimal solutions for the model, provided that enough time is given. If the optimization is aborted earlier, they may still yield useful approximate solutions together with quality guarantees (bounds).

MIP solvers are mainly based on branch-and-bound in combination with advanced linear programming techniques such as the cutting plane method utilizing a variety of generic and special purpose types of cuts, i.e., valid inequalities for strengthening a model, in order to derive tight bounds for subproblems early in the process. In addition, advanced preprocessing techniques and various kinds of heuristics and even concepts of metaheuristics also play crucial roles. Due to the latter, such solvers can be seen already as advanced hybrid optimization systems themselves.

---

[2] http://www-01.ibm.com/software/commerce/optimization/cplex-optimizer

[3] http://www.gurobi.com

[4] http://scip.zib.de

[5] http://www.solver.com/xpress-solver-engine

But even today's best MIP solvers have their limits due to the intrinsic difficulty of many hard practical problems. Frequently only small to maybe medium-sized instances can be effectively solved with practically sufficient quality. One has then only the options to (1) come up with a different, more suitable MIP formulation, (2) simplify the model by neglecting or only approximating certain aspects from the real problem, or (3) turn to other methods like more advanced mathematical programming techniques—or (meta)heuristics.

A thorough introduction to MIP is out of scope of this book, as MIP and more generally mathematical programming are huge research areas on their own. Excellent classic textbooks about solving combinatorial optimization problems by MIP include Bertsimas and Tsitsiklis [17], Wolsey [299], and Nemhauser and Wolsey [203]. We will, however, cover some basics of MIP modeling and some advanced decomposition techniques, as the area of MIP in general provides plenty of opportunities for the hybridization with metaheuristics in order to address large difficult problems more effectively.

Let us start with some basic notation. An *Integer Linear Programming* (ILP) model can be stated in the form

$$z_{\mathrm{ILP}} = \min\{c^T x \mid Ax \geq b,\ x \geq 0,\ x \in \mathbb{Z}^n\}, \tag{1.5}$$

where x is a non-negative n-dimensional integer variable vector of domain $\mathbb{Z}^n$ and $c \in \mathbb{Q}^n$ an n-dimensional vector of constant "costs". Their dot-product $c^T x$ is the *objective function* that shall be minimized. Maximization problems can be transformed into minimization problems by simply changing the sign of c. Matrix $A \in \mathbb{Q}^{m \times n}$ and the m-dimensional vector $b \in Q^m$ together define m inequality constraints. Less-than inequalities can be brought into the required greater-than-or-equal-to form by changing the sign of the corresponding coefficients, and equalities can be translated into pairs of inequalities. Thus, without loss of generality, we restrict our following considerations to minimization problems of this standard form if not explicitly stated otherwise. While in an ILP model all variables are discrete, a MIP model also allows continuous variables.

In general, MIPs do not necessarily have an optimal solution as the search space defined by the constraints might be empty or unbounded. However, we assume here a practically meaningful model for a problem with a bounded non-empty space of feasible solutions.

One of the most important concepts in MIP are *relaxations*, where some constraints of a problem are loosened or omitted. Relaxations are mostly used to obtain related simpler problems that can be solved more efficiently and yield bounds or approximate—but not necessarily feasible—solutions to the original problem. Embedded within a branch-and-bound framework, these techniques may be the basis for effective exact solution techniques, but they provide as well good starting points for heuristics.

The *linear programming* (LP) (or continuous) relaxation of the ILP (1.5) is obtained by relaxing the integrality constraints, i.e., by replacing the discrete variable domain Z^n for x by the continuous domain $\mathbb{R}^n$, yielding

$$z_{\text{LP}} = \min\{c^T x \mid Ax \geq b,\ x \geq 0,\ x \in \mathbb{R}^n\}. \tag{1.6}$$

In practice, even very large instances of such LPs can be efficiently solved by simplex-based or interior point algorithms. The solution value provides a lower bound for the original ILP, i.e., $z_{\text{LP}} \leq z_{\text{ILP}}$, since the search space of the ILP is contained within that of the LP and the objective function remains the same.

To the LP (1.6) we can further associate the *dual LP problem*

$$w_{\text{LP}} = \max\{u^T b \mid u^T A \leq c,\ u \geq 0,\ u \in \mathbb{R}^m\}, \tag{1.7}$$

with u being the m-dimensional dual variable vector. The dual of the dual LP is the original (*primal*) LP again, and the following relations are important theorems.

Strong duality: If the primal LP has a finite optimal solution with value z^*_{LP}, then its dual also has an optimal solution with same value $w^*_{\text{LP}} = z^*_{\text{LP}}$.

Weak duality: The value of every finite feasible solution to the dual problem is a lower bound for the primal LP, and each value of a finite feasible solution to the primal is an upper bound for the dual problem. As a consequence, if the dual is unbounded, the primal is infeasible and vice versa.

When considering an ILP (or MIP), we have to distinguish: A *weak dual* of ILP (1.5) is any maximization problem $w = \max\{w(u) \mid u \in S_{\text{D}}\}$ such that $w(u) \leq c^T x$ for all $x \in \{Ax \geq b,\ x \geq 0,\ x \in \mathbb{Z}^n\}$. An example of a weak dual of the ILP (1.5) is the dual (1.7) of its LP relaxation (1.6). A *strong dual* would be a weak dual that further has an optimal solution u^* with a coinciding objective value $w(u^*) = cx^*$ for an optimal solution x^* of ILP (1.5). For solving MIPs weak duals that are iteratively strengthened during the course of the optimization process are often utilized, but also in metaheuristics solutions to weak duals may provide valuable guidance.

LP-based branch-and-bound primarily utilizes the LP-relaxation to obtain dual bounds, i.e., lower bounds in case of minimization, for solving MIPs and provides the fundamental basis for the above-mentioned general purpose MIP solvers.

1.6 Constraint Programming

Constraint Programming (CP) is a technique more oriented towards *Constraint Satisfaction Problems* (CSP), i.e., problems involving complicated constraints for which already finding *any* feasible solution is difficult and either no objective function is given at all or it plays a rather secondary role only.

In contrast to MIP, CP can sometimes effectively deal with a larger variety of different constraint types and in particular also nonlinear dependencies. Similarly to

MIP, powerful generic solvers exist nowadays such as the open-source frameworks Gecode[6] and Choco[7] and the commericial IBM CPLEX CP Optimizer[8].

Generically, a CSP can be defined as a triple $P = (x, D, C)$ with $x = (x_1, \ldots, x_n)$ being an n-dimensional variable vector, $D = (D_1, \ldots, D_n)$ a corresponding vector of domains such that $x_i \in D_i$, $i = 1, \ldots, n$, holds, and $C = (C_1, \ldots, C_m)$ a set of constraints, where each constraint C_j, $j = 1, \ldots, m$, is defined by a relation R_{S_j} on the variables in the constraint's scope $S_i \subseteq \{x_1, \ldots, x_n\}$. The relation R_{S_j} can further be defined as a subset of the Cartesian product of the domains of the variables in S_i, i.e., it restricts the values that the variables can simultaneously take. The goal is to find a feasible solution to the CSP P, i.e., an instantiation of x satisfying all the constraints C.

As with mathematical programming, CP comprises a large number of very specific techniques, for which a comprehensive overview would clearly exceed the scope of this book. We refer in particular to the textbook [259] for an in-depth presentation of CP, but also to [134] when it comes to combining CP methods with other CO techniques.

There are two main principles involved in classic CP: *inference* and *search*. The latter is closely related to what we already discussed in Section 1.3, i.e., tree search methods form an important basis. We therefore focus on inference in the following.

Inference is used for reducing a given CSP to achieve some form of *local consistency* by applying a logical deduction process called *constraint propagation*. Let us start with a simple example: Assume variable x_1 has domain $D_1 = \{1,2,3,4\}$ and there exists a constraint $x_1 \leq 3$. Then, obviously the domain of x_1 can be reduced to $D_1 = \{1,2,3\}$ and the constraint can be discarded. Applying this kind of domain reduction for all unary constraints, so-called *node consistency* is achieved, which is the simplest form of local consistency.

A more advanced kind of consistency is *arc consistency*: A variable x_1 of a CSP is arc-consistent with another variable x_2 with respect to a given binary constraint on these variables if each of the admissible values in D_1 is consistent with some admissible value in D_2, i.e., if for any $x_1 \in D_1$ there exists at least one value in D_2 such that the binary constraint between x_1 and x_2 is fulfilled. As an example assume $D_1 = D_2 = \{1,2,3\}$ and the constraint $x_1 < x_2$. This CSP is not arc consistent as for $x_1 = 3$ no valid assignment for x_2 exists. Constraint propagation therefore removes the value 3 from D_1. More generally, any CSP involving only binary constraints can be made arc consistent by iteratively considering all pairs of variables for which binary constraints exist and removing values from the domain of a variable for which no valid assignment for the second variable exists. This process is continued until no domain can be further reduced.

Path consistency further generalizes arc consistency by considering pairs of variables and their consistency w.r.t. a third variable and all binary constraints on these three variables. Furthermore, arc and path consistency can be generalized to non-

---

[6] http://www.gecode.org

[7] http://choco-solver.org

[8] http://www-01.ibm.com/software/commerce/optimization/cplex-cp-optimizer

binary constraints using tuples of variables instead of a single variable or a pair of variables.

Clever constraint propagation algorithms are known for achieving these and other, more specific forms of consistency. Constraint propagation frequently provides a means to dramatically reduce the effective search space of a highly constrained problem in a time-efficient manner. Embedded in a tree search, the number of search tree nodes might be kept small enough so that sometimes even large instances of CSPs can be solved well in practice.

1.7 Why Hybridization?

After presenting some of the most important metaheuristics and exact techniques, let us briefly treat the question of why the combination of metaheuristics with other techniques for optimization—and in particular with the above sketched exact techniques—might be fruitful. Over recent years, quite an impressive number of algorithms were reported that do not purely follow the paradigm of a single traditional metaheuristic. On the contrary, they combine various algorithmic components, often originating from algorithms of other research areas on optimization. These approaches are commonly referred to as *hybrid metaheuristics*.

The main motivation behind the hybridization of different algorithms is to exploit the complementary character of different optimization strategies, that is, hybrids are believed to benefit from *synergy*. In fact, choosing an adequate combination of complementary algorithmic concepts can be the key for achieving top performance in solving many hard optimization problems. Unfortunately, developing an effective hybrid approach is in general a difficult task which requires expertise from different areas of optimization. Moreover, the literature shows that it is nontrivial to generalize, that is, a certain hybrid might work well for specific problems, but it might perform poorly for others. Nevertheless, there are hybridization types that have been shown to be successful for many applications. They may serve as a guidance for new developments.

The work on hybrid algorithms is relatively recent. During the first two decades of research on metaheuristics, different research communities working on metaheuristic techniques co-existed without much interaction, neither among themselves nor with, for example, the Operations Research community. This was justified by the fact that initially pure metaheuristics had considerable success: For many problems they quickly became state-of-the-art algorithms. However, the attempt to be different to traditional Operations Research has led to a pernicious disregard of valuable optimization expertise that was gathered over the years. Only when it became clear that pure metaheuristics had reached their limits, more researchers turned towards the development of hybrid metaheuristics.

The lack of a precise definition of the term *hybrid metaheuristics* has sometimes been subject to criticism. Note, however, that the relatively open nature of this expression can be helpful, as strict borderlines between related fields of research are often a hindrance for creative thinking and the exploration of new research directions. Conferences and workshops such as *CPAIOR* [219, 287, 175], *Hybrid Metaheuristics* [23, 24], and *Matheuristics* [187, 129, 80, 188] document the growing popularity of this line of research. Moreover, the first book specifically devoted to hybrid metaheuristics was published in 2008 [34]. The interested reader can find reviews on hybrid metaheuristics in [67, 83, 243, 34, 247, 188].

Finally, we would like to emphasize that this book covers the area of hybrid metaheuristics for single-objective combinatorial optimization problems. Readers interested in recent developments concerning hybrid metaheuristics for multiobjective optimization are referred to a survey specifically devoted to this topic [86]. Concerning the very active field of (hybrid) metaheuristics for continuous—i.e., real parameter—optimization, readers may find a good starting point in recent papers published in a dedicated journal special issue [196]. Especially in the fields of evolutionary algorithms and swarm intelligence, the use of pure as well as hybrid metaheuristics for continuous optimization has already a quite long tradition [237, 200, 87]. For an overview on parallel hybrid metaheuristics we recommend [69, 280]. Finally, it is also important to mention the availability of several currently available software frameworks that support the implementation of hybrid metaheuristics for single- and multiobjective optimization, both in a sequential and in a parallel way. One of the most powerful ones is PARADISEO [46].

1.8 Further Outline of the Book

For the following five chapters we selected five more generally applicable hybridization strategies. Some of them—as, for example, decoder-based approaches and Large Neighborhood Search presented in Chapters 2 and 4, respectively—have proven to be successful on a large variety of applications. Others, such as algorithms based on problem instance reduction (Chapter 3) and complete solution archives (Chapter 6), are rather recent, but highly promising for certain classes of problems. In the following we briefly outline the content of each of these chapters.

The techniques presented in Chapter 2 deal with representing solutions to a CO problem in indirect and/or incomplete ways. Hereby, the actual solutions to the tackled problem are obtained by the application of a so-called decoder algorithm, which might be anything from a heuristic to a complete technique. The benefits of an indirect solution representation are to be found in the fact that a complex search space, possibly with constraints that are difficult to handle, is transformed into a search space in which standard metaheuristic operators can more easily be applied. Furthermore, the decoder might be used to solve certain aspects of the whole problem, and the metaheuristic framework may then act on a higher level. The example of the Generalized Minimum Spanning Tree problem is used to show different ways of representing solutions in an indirect way.

Chapter 3 discusses a hybridization scheme based on the following basic observation. General MIP solvers, such as CPLEX and GUROBI, are often effective up to a certain, problem-specific, instance size, but beyond this size their performance degrades rapidly. When a given problem instance is too large to be directly solved by a MIP solver, it might be possible to reduce the problem instance in a clever way such that the resulting reduced problem instance contains high-quality—or even optimal—solutions to the original problem instance. The reduced problem instance must, of course, be small enough in order to be effectively tackled by a MIP solver. In the context of the application to the Minimum String Partition Problem we present an algorithm that makes use of this principle.

Large Neighborhood Search considered in Chapter 4 is probably one of the most well-known, and most successful, hybrid techniques to date. Given an incumbent solution to the tackled problem instance, the algorithm follows the general idea of first destroying the incumbent solution partially, and then applying, for example, a MIP solver to find the best valid solution that contains the given partial solution. Note the relationship to the iterated greedy metaheuristic outlined in Section 1.2.2. Here, however, the re-generation step is done in a more sophisticated way, such as applying a MIP solver to a corresponding ILP model for the original problem, in which the part of the variables corresponding to the given partial solution has already fixed values. The motivating aspect for this idea is the same as that described in Chapter 3. Even though it might be unfeasible to apply a MIP solver to the original problem instance, the solver might be able to effectively solve the ILP model in which only a part of the variables remains open. The application of this scheme is demonstrated in the context of the Minimum Weight Dominating Set and the Generalized Minimum Spanning Tree problems.

Chapter 5 describes a hybridization technique which is especially suited for metaheuristics primarily based on the construction of solutions, as opposed to metaheuristics based on neighborhood search. In such metaheuristics, solution constructions are most frequently performed sequentially and independently from each other. Moreover, greedy information is usually used in a probabilistic way as the only source of knowledge on the tackled problem (instance). In contrast, more classic tree search based methods, such as the heuristic branch-and-bound-derivative beam search, make use of two different types of problem knowledge. In addition to greedy information, they exploit bounding information in order to prune branches of the search tree and to differentiate between partial solutions on the same level of the search tree. In other words, these methods can be seen as generating solutions in a (quasi-)parallel, non-independent way. The general idea of the approach presented in this chapter is to improve metaheuristics based on solution construction by incorporating bounding information and the parallel and non-independent construction of solutions. This is illustrated in the context of the Multidimensional Knapsack Problem.

The last chapter dedicated to a specific hybridization principle—Chapter 6—shows how a metaheuristic might profit from the incorporation of basic principles of branch-and-bound. More specifically, we extend an evolutionary algorithm by a complete solution archive that stores all considered candidate solutions organized along the principles of a branch-and-bound tree. The approach is particularly appealing for problems with expensive evaluation functions or a metaheuristic applying indirect or incomplete solution representations in combination with non-trivial decoders. Besides just avoiding re-evaluations of already considered solutions, a fundamental feature of the solution archive discussed in this chapter is its ability to efficiently transform duplicates into typically similar but guaranteed new solution candidates. From a theoretical point of view, the metaheuristic is turned into a complete enumerative method without revisits, which is in principle able to stop in limited time with a proven optimal solution. Furthermore, the approach can be extended by calculating bounds for partial solutions possibly allowing us to prune larger parts of the search space. In this way the solution archive enhanced metaheuristic can also be interpreted as a branch-and-bound optimization process guided by principles of the metaheuristic search. The presented example application concerns again the Generalized Minimum Spanning Tree problem.

Finally, the last chapter references some further more general hybridization principles that are not dealt with in this book in more detail, such as hybrid approaches based on mathematical programming decomposition techniques, and it concludes this book.

Chapter 2
Incomplete Solution Representations and Decoders

Representing candidate solutions in a metaheuristic in an indirect way and using a decoding algorithm for obtaining corresponding actual solutions is a commonly applied technique to transform a more complex search space, possibly with constraints that are difficult to handle, into one where standard local search neighborhoods can be more easily applied. The decoding algorithm used here may also be a more advanced, "intelligent" procedure that solves part of the whole problem. This leads us further to *incomplete* solution representations, where a metaheuristic essentially acts on only a subset of the decision variables, while the decoder augments the missing parts in an optimal or reasonably good way. We study this general approach by considering the *Generalized Minimum Spanning Tree* (GMST) problem as an example and investigate two different decoder-based variable neighborhood search approaches relying on complementary incomplete representations and respective efficient decoders. Ultimately, a combined approach is shown to perform best.

This chapter is organized as follows. After discussing the general ideas of decoder-based metaheuristics and incomplete solution representations, the Generalized Minimum Spanning Tree problem is introduced. It essentially extends the classic minimum spanning tree problem by considering a clustered graph and the requirement that exactly one node from each cluster needs to be connected. We then consider a VNS approach in which solutions are represented incompletely only by the selection of nodes that are to be connected; corresponding edges are identified by a decoder corresponding to a classic minimum spanning tree algorithm. As an alternative, we then study a dual approach, in which the VNS decides which clusters are to be directly connected and a decoding procedure based on DP is used for identifying optimal node selections. Both strategies are ultimately combined into a VNS making use of both representations and the respective neighborhood structures. Experimental comparisons indicate the advantages of these decoder-based approaches and the dual-representation approach in particular.

2.1 General Idea

Indirect solution representations are known especially well from genetic algorithms [116]: There, one frequently distinguishes between the *genotype*, which is the solution in an encoded form, and the *phenotype*, corresponding to the actual solution in a natural representation. One main goal achieved by this differentiation is that standard variation operations like one-point, two-point, or uniform crossover and position-wise mutation can be applied—the actual genetic algorithm with its operators to derive new candidate solutions can stay more or less generic; i.e., the problem specificities are encapsulated in a decoding procedure that transforms a genotype into its corresponding phenotype. Also other metaheuristics occasionally make use of such indirect representations, especially when the original search space is complex due to difficult-to-handle constraints and simple neighborhood structures might yield infeasible solutions too frequently.

Such decoder-based approaches, however, are also an opportunity to exploit some other efficient optimization technique within a metaheuristic search. As long as the decoding procedure is not too slow, it may make perfect sense to utilize an "intelligent" algorithm for this purpose which can take care of certain aspects or parts of the problem. Solutions may then only be *incompletely* (and indirectly) represented within the metaheuristic and the decoder takes care of possibly transforming encoded solutions but especially augmenting missing parts. In general, decoders can be any kind of algorithms, including very problem-specific heuristics or exact procedures, MIP approaches, but especially also DP and CP. The only basic requirement is that they are sufficiently fast to still allow the metaheuristic framework to perform a reasonable number of iterations.

2.2 Application to the Generalized Minimum Spanning Tree Problem

The GMST is an extension of the classic Minimum Spanning Tree (MST) problem and can be defined as follows. Given is an undirected weighted complete graph $G = (V,E,c)$ with node set V, edge set $E = V \times V$, and edge costs $c : E \to \mathbb{R}^+$. For the cost of the edge $(u,v) \in E$ connecting nodes u and v, we will also write $c(u,v)$ instead of $c((u,v))$, with $c(u,v) = c(v,u)$. Node set V is partitioned into r pairwise disjoint clusters $V_1,\dots,V_r$ so that $\bigcup_{i=1,\dots,r} V_i = V$ and $V_i \cap V_j = \emptyset$, $\forall i,j = 1,\dots,r$, $i \neq j$. Let $d_i = |V_i|$ denote the number of nodes in cluster i.

A (classic) spanning tree of a graph is a cycle-free subgraph connecting all nodes. A feasible solution to the GMST problem on G is a cycle-free subgraph $S = (P,T)$ spanning exactly one (arbitrary) node from each cluster V_i, i.e., $P = \{p_i \mid p_i \in V_i,\ i = 1,\dots,r\}$ and $T \subset P \times P$, see Figure 2.1. The cost of such a GMST is the sum of the costs of its edges, i.e., $c(T) = \sum_{(u,v)\in T} c(u,v)$, and the objective is to identify a solution with minimum cost.

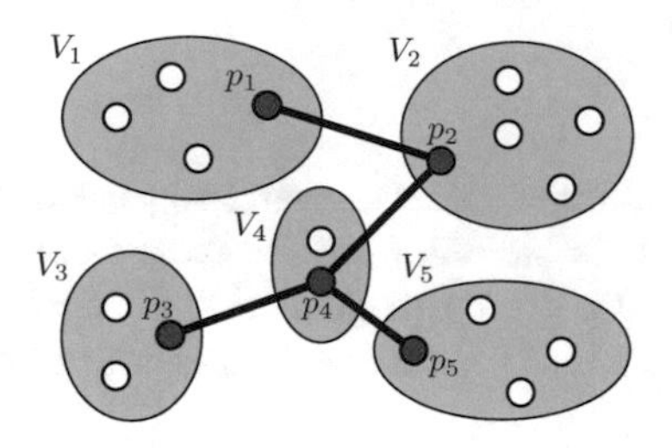

Fig. 2.1 Exemplary GMST solution (from [142])

The GMST problem was first mentioned by Myung et al. [202], who showed it to be strongly NP-hard in general by reduction from the node cover problem. Just when each cluster is a singleton, i.e. $|V_i| = 1, \ \forall i = 1,\ldots,r$, the problem obviously reduces to the classic MST problem and can be solved efficiently in polynomial time, e.g., by Kruskal's or Prim's classic MST algorithms [167, 238].

To get an idea of where this problem appears in practice, consider the design of some communication network connecting several *Local Area Networks* (LANs), e.g., in different buildings. Each LAN has several potential access points from outside, corresponding to the nodes forming one cluster in the GMST. A minimum cost tree connecting one node from each LAN is sought. For a more detailed overview on the GMST problem and its applications, we refer to [97].

Variants of the GMST include the less restrictive At-Least GMST problem in which more than one node is allowed to be connected per cluster [143, 81]. Furthermore, the GMST and At-Least GMST problems can be considered a special case of the Group Steiner Problem [253], in which groups of nodes that are not necessarily disjoint are considered instead of clusters. Again, a tree spanning at least one node from each group is sought.

Existing approaches to solve the GMST include diverse MIP formulations proposed in [202, 96, 232] and a more complex branch-and-cut algorithm [97]. These exact approaches are, however, in practice limited to small and medium-sized instances. Depending on the structure and specific parameters, instances with up to about 200 nodes can be solved to proven optimality by these methods. For constant cluster size d Pop et al. [233] describe a $2d$ approximation algorithm, and a polynomial time approximation scheme for the special case of grid-clustered graphs was proposed by Feremans et al. [96].

Concerning metaheuristics to approach larger instances in practice, Ghosh [109] investigated two tabu search algorithms and several VNS variants, which are based on similar principles as those that we will study in the first approach considered in this chapter. Golden et al. [117] investigated a genetic algorithm for the GMST problem. Hu, Leitner, and Raidl [142] proposed advanced VNS variants on which the rest of this chapter mainly relies; for more details in this respect also see Hu's PhD thesis [138]. Öncan et al. [209] described another advanced tabu search, and Jiang and Chen [146] suggested a metaheuristic called the *Dynamic Candidate Set Algorithm* for the GMST.

2.2.1 *Initial Solutions*

For obtaining an initial solution quickly, several constructive heuristics can be used. Ghosh [109] describes a simple *Minimum Distance Heuristic* (MDH) and Golden et al. [117] compared three greedy methods that are derived in rather straightforward ways from Kruskal's [167], Prim's [238] and Sollin's classic MST algorithms, respectively.

In our approaches, we consider MDH and an iterated variant of the Kruskal-based construction method as detailed in [142]. As both methods are fast, yield different solutions most of the time, and are on average similarly effective, we perform both and take the better solution as the initial one for the successive VNS. In the following we summarize the principles of the two construction heuristics.

Minimum Distance Heuristic (MDH)

This heuristic first selects for each cluster the node with the lowest sum of edge costs to all nodes in other clusters. Then Kruskal's MST algorithm is applied to the subgraph induced by these nodes. As choosing the nodes requires checking all edge costs and the MST can be determined in $O(r^2 \log r)$ time, MDH runs in $O(|V|^2 + r^2 \log r)$ time.

Iterated Kruskal-Based Heuristic (IKH)

Here, all edges E are considered in non-decreasing cost order; ties are broken randomly. An edge $(u, v) \in E$ is added to the solution's edge set T if and only if it does not introduce a cycle and no other edge has already been selected for T connecting some other node(s) from the clusters to which u and v belong to. By always fixing one specific node $u \in V$ to be connected in the solution and iterating this construction process, different solutions are obtained and a best one is chosen as final result. The overall time complexity of this method is $O(|V||E|)$ as the sorting of edges only needs to be done once.

2.2.2 *Node Set Based VNS (NSB-VNS)*

For an overview on the general principles of VNS see Section 1.2.6. As already observed, the GMST problem becomes easy when the nodes to be connected from all the clusters are given, as in this case the problem reduces to the classic MST problem.

Following the idea of incomplete solution representations and applying decoders, it is therefore straightforward to define a metaheuristic that concentrates on finding a best-suited selection of nodes P and leaves the determination of correspondingly best-suited edges to an efficient MST algorithm. We call this way of representing a solution *Node Set Based* (NSB) representation. Here we utilize Prim's MST algorithm as decoder, as with its time complexity of $O(r^2)$ it can be implemented more efficiently for complete graphs than Kruskal's algorithm. A solution in our *Node Set Based VNS* (NSB-VNS) is thus solely represented by a vector of nodes $p = (p_1, \ldots, p_r) \in V_1 \times \ldots \times V_r$.

Node Exchange Neighborhood

For locally improving solutions, the following *Node Exchange Neighborhood* (NEN) is most intuitive [109]: For a current solution represented by p it includes all node vectors p' that differ in exactly one selected node, i.e., where one node p_i is replaced by a different node p'_i of the same cluster. NEN thus consists of $\sum_{i=1}^{r}(|V_i| - 1) = O(|V|)$ different node vectors, and decoding and evaluating each neighbor independently can be accomplished in time $O(|V| \cdot r^2)$ in order to obtain a best neighboring solution.

This computational effort for searching a complete neighborhood can, however, be sped up significantly by applying an *incremental evaluation scheme*. Incremental evaluation means that the objective value—and here also the complete solution—is not determined from scratch for each neighbor but is derived from the original solution by only considering the changes. In our case, p_i is removed with all its incident edges from the initial tree $S = (P, T)$ represented by p. Denoting the degree of node p_i in tree $S = (P, T)$ by $\deg(p_i)$, this implies that $\deg(p_i)$ connected components remain. Observe that a minimum cost tree $S' = (P', T')$ for the solution represented by $p' = (p_1, \ldots, p_{i-1}, p'_i, p_{i+1}, \ldots, p_r)$ will not require any different edges in these connected components, as each component is cost-minimal w.r.t. its nodes. Therefore, we can build on these components and reconnect them with the new node p'_i in a cost-minimal way. New edges are only necessary between nodes of different components and/or p'_i, and only the shortest edges connecting any pair of components need to be considered. Thus, the edges T' must be a subset of

- $T \setminus \{(p_i, u) \mid u \in V \setminus \{p_i\}\}$,
- edges (p'_i, p_j) with $j = 1, \ldots, r \wedge j \neq i$, and
- the least-cost edges between any pair of the remaining connected components of S after removing p_i with its incident edges.

To compute S', we therefore have to calculate the MST of a graph with $(r - \deg(p_i) - 1) + (r - 1) + \deg(p_i) \cdot (\deg(p_i) + 1)/2 = O(r + \deg(p_i)^2)$ edges only. Unfortunately, this incremental evaluation does not change the worst-case time complexity, because identifying the shortest edges between any pair of components may require time $O(r^2)$ when $\deg(p_i) = O(r)$. However, in most practical cases it is substantially faster to compute these shortest edges and to apply Kruskal's MST algorithm on the resulting thin graph. Especially when replacing a leaf node of the initial tree S, we only get a single component plus the new node and the incremental evaluation's benefits are obviously largest.

General k-Node Exchange Neighborhoods

The NEN can be generalized by considering a simultaneous replacement of $k > 1$ nodes. The neighborhood size and computational complexity, however, also grow rapidly with increasing k. k-NEN has in general $O(|V|^k)$ neighbors, and an independent decoding and evaluation of all these node vectors requires time $O(|V|^k \cdot r^{2k})$. For larger instances, unfortunately, this computational complexity soon becomes too large. Even incremental evaluation techniques are not able to reduce the time-effort enough in order to make a complete neighborhood search meaningful. Note that by removing more than one node from a current tree, the number of obtained components also increases roughly proportional to k in the expected case. Actually, already 2-NEN turns out to be practically too demanding to be completely searched for instances with more than a few hundred nodes.

For diversifying the heuristic search in the context of our VNS, however, more general k-NEN with $k \geq 2$ make perfect sense, as only single random solutions are sampled per shaking iteration.

Restricted Two-Nodes Exchange Neighborhood

Sometimes, when a neighborhood is too large to be searched completely, considering just a promising subset of neighbors by imposing some restriction can be very meaningful. In the context of the GMST, Hu et al. [142] came up with the following *Restricted Two-Nodes Exchange Neighborhood* (R2NEN), which we adopt here: Only pairs of nodes that are adjacent in the current solution S are simultaneously considered. Assuming the current solution is already locally optimal w.r.t. NEN, it appears on average more likely that further improvements can be achieved by considering changes in such currently directly connected clusters for which a good selection of nodes might be stronger correlated than when considering some other arbitrary distant clusters. Supposing the clusters are of similar size, a complete evaluation of R2NEN can be done in only $O(|V| \cdot r^2)$ time.

To finally wrap up the characteristics of the NSB-VNS:

- Initially, MDH and IKH are applied and a best solution is kept as a starting solution for the VNS.
- The VNS framework as already shown in Algorithm 9 in Section 1.2.6 is used.
- Each candidate solution undergoes local improvement in a VND fashion (see Section 1.1.2.1), in which NEN followed by R2NEN are efficiently searched by incremental evaluation; a best improvement step function is applied and the local search continues until a solution that is locally optimal w.r.t. both neighborhood structures is reached.
- Shaking is performed by applying a random move in the general k-NEN, with $k = 3, \ldots, k_{\max}$. (Shaking in 2-NEN is not considered, as the successive VND would mostly lead back to the same local optimum.)

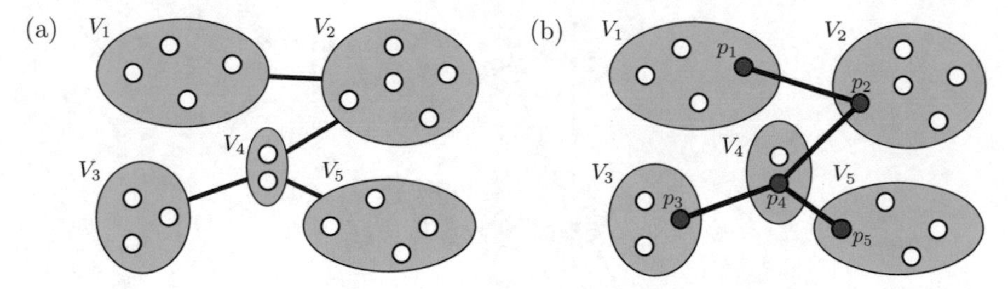

Fig. 2.2 (a) A global spanning tree S^g with edge set T^g and (b) a corresponding cost-minimal generalized spanning tree from $S(T^g)$ when considering Euclidean distances as edge costs (from [142])

2.2.3 Global Edge Set Based VNS (GESB-VNS)

A complementary, also incomplete way to represent a solution to the GMST problem follows the idea of specifying which clusters are directly connected without considering specific nodes. For this purpose, we define for a GMST instance the *global graph* $G^g = (V^g, E^g)$ consisting of nodes corresponding to clusters in G, i.e. $V^g = \{V_1, V_2, \ldots, V_r\}$, and edge set $E^g = V^g \times V^g$.

We now represent a candidate solution for the GMST problem in the metaheuristic search by a global edge set $T^g \subseteq E^g$ forming a spanning tree $S^g = (V^g, T^g)$ on the global graph G^g and call this encoding *Global Edge Set Based* (GESB) representation, see Figure 2.2. Such a global edge set T^g can be seen to refer to the subset $S(T^g)$ of all generalized spanning trees S in G containing for each edge $(V_i, V_j) \in T^g$ a corresponding edge $(u, v) \in E$ with $u \in V_i \land v \in V_j \land i \neq j$. This subset of GMST solutions $S(T^g)$ is in general exponentially large with respect to the number of nodes $|V|$. However, by using a dynamic programming procedure originally suggested by Pop [232] and described in the next paragraphs, we can always efficiently determine a cost-minimal solution from $S(T^g)$ for any given T^g. Thus, this DP procedure is used to augment the incomplete solution T^g by additionally selecting a best-suited node p_i from each cluster V_i, $i = 1, \ldots, r$.

Dynamic Programming for Decoding Global Spanning Trees

For a general overview on the principles of DP, see Section 1.4. Here, we first root the global spanning tree S^g at an arbitrary cluster $V_{\mathrm{root}} \in V^g$ and direct all edges to obtain an outgoing arborescence, i.e., directed paths from the root towards all leaves. Then we traverse this tree in a depth-first fashion calculating for each cluster $V_i \in V^g$ and each node $v \in V_i$ the minimum costs for the subtree rooted in V_i when v is the node to be connected from V_i. Minimum costs of a subtree are determined by the recursion

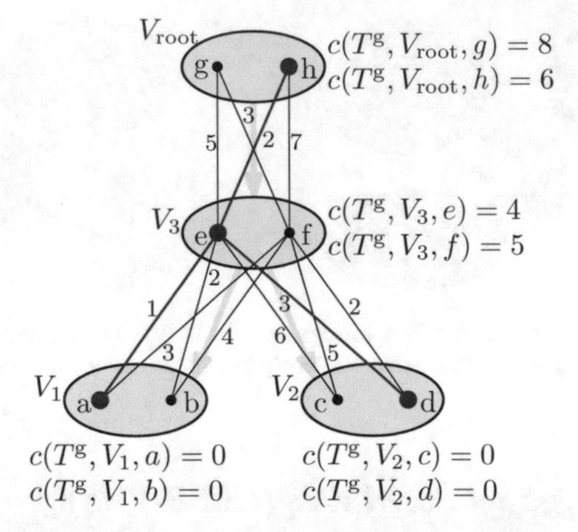

Fig. 2.3 DP for the GESB representation: Determining minimum cost values for each cluster and node; the tree's overall minimum costs are $c(T^{\text{g}}, V_{\text{root}}, h) = 6$, and the finally chosen nodes are printed in larger size (from [142])

$$c(T^{\text{g}}, V_i, v) = \begin{cases} 0 & : \text{ if } V_i \text{ is a leaf of } S^{\text{g}} \\ \sum\limits_{V_l \in \text{succ}(V_i)} \min\limits_{u \in V_l} \{c(v,u) + c(T^{\text{g}}, V_l, u)\} & : \text{ else,} \end{cases}$$

where $\text{succ}(V_i)$ denotes the set of all successors of V_i in S^{g}. After having determined the minimum costs for the whole tree, the nodes to be used can be derived in a top-down fashion by fixing for each cluster $V_i \in V^{\text{g}}$ a node $p_i \in V_i$ yielding minimum costs. This DP procedure is illustrated in Figure 2.3. It requires in general in the worst-case $O(|V|^2)$ time.

Global Edge Exchange Neighborhood

For defining a suitable neighborhood structure, we have to take care that a feasible global edge set T^{g} must always form a spanning tree of the global graph G^{g}. Thus we may not arbitrarily add, remove or exchange edges. The *Global Edge Exchange Neighborhood* (GEEN) contains any feasible spanning tree differing from T^{g} by exactly one edge. As any spanning tree of G^{g} always has $|r|-1$ edges, there are this many possibilities to remove a single edge, yielding two disconnected components. These components are then connected again by a different edge, yielding a total of $O(r^3)$ neighboring spanning trees. If we determine the best generalized spanning tree in this neighborhood by evaluating all possibilities and naively performing the whole DP for each global spanning tree, the total time complexity is $O(|V|^2 \cdot r^3)$.

For a more efficient evaluation of all neighbors, we perform the whole DP only once at the beginning, store all costs $c(T^{\text{g}}, V_i, v)$, $\forall i = 1, \ldots, r,\ v \in V_i$, and incrementally update our data for each considered exchange. According to the recursive

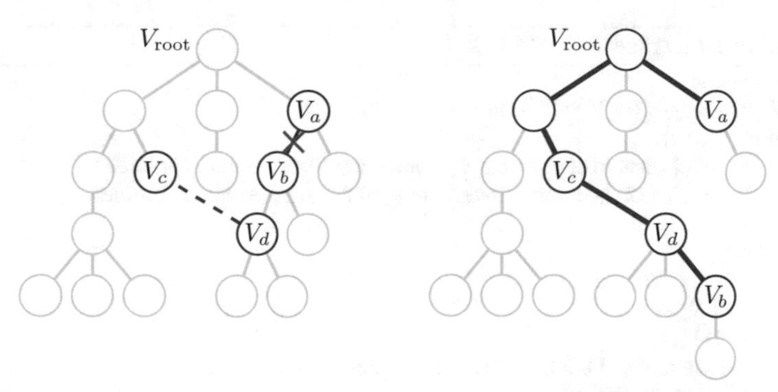

Fig. 2.4 Incremental DP: After deleting (V_a, V_b) and inserting (V_c, V_d), only the clusters on the paths from V_a to V_{root} and V_b to V_{root} need to be reconsidered (from [142])

definition of the DP, we only need to recalculate the values of a cluster V_i if its set of successors or their costs change.

Figure 2.4 illustrates the move from one global spanning tree to a neighboring one by exchanging global edge (V_a, V_b) with (V_c, V_d). When removing (V_a, V_b), the subtree rooted at V_b is disconnected, hence V_a loses a child and V_a as well as all its predecessors must be updated. Before we add (V_c, V_d), we first need to consider the isolated subtree. If $V_d \neq V_b$, we have to re-root the subtree at cluster V_d. Hereby, the old root V_b loses a child. All other clusters which get new children or lose children are on the path from V_b up to V_d, and they must be re-evaluated. Otherwise, if $V_d = V_b$, nothing changes within the subtree. When adding the edge (V_c, V_d), V_c gets a new successor and therefore must be updated together with all its predecessors on the path up to the root. In conclusion, whenever we replace an edge (V_a, V_b) by another edge (V_c, V_d), it is enough to update the costs of V_a, V_b, and all their predecessors on their paths up to the root of the new global tree.

If the tree is not degenerated, its height is $O(\log r)$, and we only need to update $O(\log r)$ clusters of G^{g}. Suppose each of them contains no more than $d_{\max}$ nodes and has at most $s_{\max}$ successors. The time for updating the costs of a single cluster V_i is then in $O(d_{\max}^2 \cdot s_{\max})$, and the whole procedure runs in time $O(d_{\max}^2 \cdot s_{\max} \cdot \log r)$. The incremental evaluation is therefore much faster than the complete evaluation with its time complexity of $O(|V|^2)$ as long as the trees are not degenerated. An additional improvement is to check if an update actually changes the costs of a cluster. If this is not the case, one can skip the update of the cluster's predecessors as long as they are not affected in some other way.

To examine the whole neighborhood of a current solution by using the improved method described above, it is a good idea to choose a processing order that further supports incremental evaluation. Algorithm 16 shows how this is done in detail.

Removing an edge (V_i, V_j) splits our rooted tree into two components: K_1^{g} containing V_i and K_2^{g} containing V_j. The algorithm iterates through all clusters $V_k \in K_1^{\text{g}}$ and makes them root. Each of these clusters is iteratively connected to every cluster of K_2^{g} in the inner loop. The advantage of this calculation order is that none of the

Algorithm 16 Exploration of GEEN

```
given: solution S
for all global edges (V_i, V_j) ∈ T^g do
    remove (V_i, V_j)
    M_1 = list of clusters in component K_1^g containing V_i, traversed in preorder
    M_2 = list of clusters in component K_2^g containing V_j, traversed in preorder
    for all V_k ∈ M_1 do
        root K_1^g at V_k
        for all V_l ∈ M_2 do
            root K_2^g at V_l
            add (V_k, V_l)
            use incremental DP to determine complete solution
                and its objective value
            if current solution better than best then
                save current solution as best
            end if
            remove (V_k, V_l)
        end for
    end for
end for
return best solution
```

clusters in K_1^g except its root V_k has to be updated more than once, because global edges are only added between the roots of K_1^g and K_2^g. Processing clusters in preorder has another benefit: Typically, relatively few clusters have to be updated when re-rooting either K_1^g or K_2^g.

As with NEN, also GEEN might in principle be generalized to a k-GEEN variant in which $k \geq 2$ edges are exchanged. While such larger neighborhoods may make sense with respect to shaking in the VNS, practical experiments indicated that the benefits of a systematic best or first improvement search are quite limited while the computational effort increases substantially. Even a restricted variant of 2-GEEN in the spirit of R2NEN appears to be too expensive for being applied to large instances, as in practice—at least on the test instances we consider in the following Section 2.2.5—searching GEEN takes already much more time than searching NEN.

In summary, GESB-VNS works as follows:

- As with NSB-VNS, MDH and IKH are initially applied and a best solution is kept as starting solution for the VNS.
- The VNS framework as already shown in Algorithm 9 in Section 1.2.6 is used.
- Each candidate solution undergoes a local search using GEEN, in a best improvement manner until a local optimum is reached.
- Shaking is performed by applying a random move in the general k-GEEN, with $k = 2, \dots, k_{\max}$.

2.2.4 Combined VNS (COMB-VNS)

The two VNS variants presented above represent candidate solutions in *dual ways*. While NSB-VNS encodes and primarily works on the selected nodes and uses a MST algorithm for augmenting each candidate solution with the best-suited edges, GESB-VNS primarily works on global edges and augments each candidate solution with the corresponding optimal node selection.

Due to these opposite approaches, the two VNS variants work on very different search spaces. Considering the main ideas from VNS, a naturally arising question is, could another VNS variant that makes use of both representations and all the respectively presented neighborhood structures perform even better than both NSB-VNS and GESB-VNS. We call this third variant *Combined VNS* (COMB-VNS). In more detail it is composed as follows:

- Again MDH and IKH are initially applied and a best solution is kept as a starting solution.
- Again, the standard VNS framework as shown in Algorithm 9 is used.
- Each candidate solution undergoes local improvement by an embedded VND making use of NEN, GEEN, and R2NEN in this sequence. This order was determined according to the computational complexity of evaluating the neighborhoods, i.e., the fastest neighborhood search is applied first. Again, the best improvement strategy is applied and the VND continues until a locally optimal solution with respect to all these neighborhoods is obtained.
- Shaking is performed by applying both, random moves in the generalized NEN and the generalized GEEN structures. More specifically, a move in the k-th shaking neighborhood corresponds to $k+1$ random NEN moves followed by k random GEEN moves, with $k = 2, \ldots, k_{\max}$. This combination, which puts more emphasis on diversity than performing only either random NEN moves or random GEEN moves has shown to yield good results in preliminary experiments.

2.2.5 Results

In the following we present a computational comparison of NSB-VNS, GESB-VNS, and COMB-VNS and also investigate the individual contributions of the different neighborhood structures. A more comprehensive computational study was done by Hu, Leitner, and Raidl [142], from where we adopt selected results.

Benchmark Instances

We consider the following GMST benchmark instance sets which were originally proposed by Ghosh [109] and were augmented with larger instances by Hu [138].

Table 2.1 Benchmark instance sets from Ghosh [109] and Hu [138]. Three different instances are considered in each set of different type and size. Each instance has a constant number of nodes per cluster

instance set	$\lvert V\rvert$	$\lvert E\rvert$	r	$\lvert V\rvert/r$	*col*	*row*	*sep*	*span*
Grouped Eucl. 125	125	7750	25	5	5	5	10	10
Grouped Eucl. 500	500	124750	100	5	10	10	10	10
Grouped Eucl. 600	600	179700	20	30	5	4	10	10
Grouped Eucl. 1280	1280	818560	64	20	8	8	10	10
Random Eucl. 250	250	31125	50	5	-	-	-	-
Random Eucl. 400	400	79800	20	20	-	-	-	-
Random Eucl. 600	600	179700	20	30	-	-	-	-
Non-Eucl. 200	200	19900	20	10	-	-	-	-
Non-Eucl. 500	500	124750	100	5	-	-	-	-
Non-Eucl. 600	600	179700	20	30	-	-	-	-

Grouped Euclidean: Squares with a side length of *span* are regularly laid out in the Euclidean plane on a grid of size $col \times row$. The number of nodes of each cluster is fixed and they are randomly distributed within the corresponding square. By changing the ratio between cluster separation *sep* and cluster span *span*, instances are generated with clusters that are from *strongly overlapping* to *widely separated.*

Random Euclidean: In these instances, all nodes are randomly scattered over a Euclidean square of size 1000×1000 and the clustering is done independently at random. Therefore, nodes from one cluster are not necessarily close to each other.

Non-Euclidean: Here, edge costs are not Euclidean anymore but independently chosen at random from the interval $[0, 1000]$.

Furthermore, the following benchmark set is taken from Feremans [95].

TSPlib-Based: Instances from the TSPlib[1] are taken and the node clustering is done as follows: In a first step r center nodes are selected by choosing the first node randomly and then always selecting as next center node a node that is farthest away from all previous centers. In the second step, all remaining nodes are assigned to a nearest center node.

Tables 2.1 and 2.2 list the sizes and characteristics of the instances used. The number of nodes ranges in the Ghosh and Hu instances from 125 to 1280, the number of clusters from 20 to 100, and the cluster size from 5 to 30. The TSPlib-based instances have up to 442 nodes and 89 clusters, with an average cluster size of 5.

[1] http://comopt.ifi.uni-heidelberg.de/software/TSPLIB95

Table 2.2 TSPlib-based instances with geographical clustering from Feremans [95]. The number of nodes varies over the clusters between $d_{\min}$ and $d_{\max}$

instance	$\lvert V\rvert$	$\lvert E\rvert$	r	$\lvert V\rvert/r$	$d_{\min}$	$d_{\max}$
gr137	137	9316	28	5	1	12
kroa150	150	11175	30	5	1	10
d198	198	19503	40	5	1	15
krob200	200	19900	40	5	1	8
gr202	202	20301	41	5	1	16
ts225	225	25200	45	5	1	9
pr226	226	25425	46	5	1	16
gil262	262	34191	53	5	1	13
pr264	264	34716	54	5	1	12
pr299	299	44551	60	5	1	11
lin318	318	50403	64	5	1	14
rd400	400	79800	80	5	1	11
fl417	417	86736	84	5	1	22
gr431	431	92665	87	5	1	62
pr439	439	96141	88	5	1	17
pcb442	442	97461	89	5	1	10

General Algorithm Setup and Testing Environment

For shaking, the maximum neighborhood size $k_{\max}$ was set to $\lceil r/2 \rceil$; this setting has been shown to be a rather robust choice in preliminary experiments. For each VNS variant, 30 independent runs were performed per instance, and in the following we consider average final objective values $\overline{c(T)}$and standard deviations $\sigma(c(T))$. All experiments were performed on a single core of a Pentium 4 PC with 2.8 GHz and 2 GB RAM. To allow a fair comparison, each run was terminated after 600 seconds CPU-time. This time limit essentially allowed all algorithms to converge reasonably well on all instances so that no substantial further improvements can be expected when one would let the algorithms continue.

Table 2.3 shows the results of NSB-VNS, GESB-VNS, and COMB-VNS for the benchmark instances from Ghosh and Hu. For each instance, the best obtained average values are highlighted with gray background.

While all three VNS variants always found the same solutions for the smallest Grouped Euclidean instances with 125 nodes, the performance differences become clear for all larger test cases: Both NSB-VNS as well as COMB-VNS yielded consistently better average final objective values than GESB-VNS, and also standard deviations are most of the time smaller. Thus, GESB-VNS on its own obviously is not a good choice.

Concerning NSB-VNS in comparison to COMB-VNS, we can see that the combined approach yields superior average objective values on 17 instances, while NSB-VNS is better in only three cases. In 10 cases, both VNS variants performed equally well. Thus, in general we can conclude that applying all neighborhood structures in a combined fashion is indeed advantageous. This is also supported by a statistical

Table 2.3 Average final objective values and standard deviations of the different VNS variants for the instance sets from Ghosh and Hu over 30 runs per instance. Three instances are considered per type and size

instance set	NSB-VNS		GESB-VNS		COMB-VNS	
	$\overline{c(T)}$	$\sigma(c(T))$	$\overline{c(T)}$	$\sigma(c(T))$	$\overline{c(T)}$	$\sigma(c(T))$
Grouped Eucl. 125	141.1	0.00	141.1	0.00	141.1	0.00
	133.8	0.00	133.8	0.00	133.8	0.00
	141.4	0.00	141.4	0.00	141.4	0.00
Grouped Eucl. 500	571.7	0.48	589.6	5.56	568.6	0.59
	586.5	1.18	601.8	5.33	581.0	1.39
	587.9	1.91	597.1	2.37	587.9	4.07
Grouped Eucl. 600	84.6	0.11	84.8	0.27	84.8	0.27
	87.9	0.02	88.3	0.24	87.9	0.05
	88.5	0.00	88.7	0.12	88.5	0.00
Grouped Eucl. 1280	323.6	1.10	327.1	4.30	321.8	2.41
	317.8	0.79	326.4	3.01	316.3	0.83
	335.6	2.16	339.9	3.87	334.3	2.13
Random Eucl. 250	2354.0	43.08	2646.6	28.49	2336.9	34.23
	2399.7	21.94	2576.2	112.15	2304.1	47.95
	2061.2	26.91	2460.6	131.82	2049.8	15.29
Random Eucl. 400	621.9	14.20	703.4	53.40	625.4	14.59
	595.3	0.00	671.4	25.82	595.3	0.14
	597.5	17.15	657.4	50.38	588.8	7.40
Random Eucl. 600	452.9	33.57	506.5	46.08	443.5	0.00
	545.1	18.05	685.0	21.40	535.2	12.20
	493.8	35.76	643.4	63.68	479.9	26.55
Non-Eucl. 200	71.6	0.00	97.3	14.23	71.6	0.02
	41.0	0.00	58.5	11.18	41.0	0.00
	52.8	0.00	56.8	0.69	52.8	0.00
Non-Eucl. 500	185.4	8.94	203.3	0.00	173.4	8.40
	159.2	7.41	237.9	1.28	154.6	6.55
	179.5	4.36	282.5	0.00	180.1	3.67
Non-Eucl. 600	16.3	2.35	47.5	9.40	15.9	2.07
	18.3	2.24	36.0	3.67	17.6	1.75
	16.2	1.97	42.0	5.87	15.1	0.22

Wilcoxon rank test over all these instances with an error probability of less than 1%. Furthermore, COMB-VNS's standard deviations are typically smaller than those of NSB-VNS, indicating a higher robustness of the combined approach.

Concerning the different types of instances, absolute differences in the average objective values between GESB-VNS and COMB-VNS (or NSB-VNS) are relatively small for Grouped Euclidean instances, significantly larger for Random Euclidean Instances and quite dramatic for Non-Euclidean Instances, indicating especially that GESB-VNS has problems in dealing with unstructured instances.

Table 2.4 shows the results for the geographically clustered TSPlib-instances. The general trends are similar to those concerning the Grouped Euclidean instances. Overall, GESB-VNS cannot compete with the other two VNS variants. However,

Table 2.4 Average final objective values and standard deviations of the VNS variants for Feremans' TSPlib-based geographically clustered instances

instance set	**NSB-VNS**		**GESB-VNS**		**COMB-VNS**	
	$\overline{c(T)}$	$\sigma(c(T))$	$\overline{c(T)}$	$\sigma(c(T))$	$\overline{c(T)}$	$\sigma(c(T))$
gr137	329.0	0.00	329.0	0.00	329.0	0.00
kroa150	9815.0	0.00	9815.0	0.00	9815.0	0.00
d198	7044.3	1.64	7044.6	2.28	7044.0	0.00
krob200	11244.0	0.00	11264.0	22.6	11244.0	0.00
gr202	242.1	0.25	242.2	0.48	242.0	0.00
ts225	62279.9	4.58	62270.7	6.35	62280.5	16.28
pr226	55515.0	0.00	55515.0	0.00	55515.0	0.00
gil262	942.9	1.36	947.0	3.63	943.2	1.63
pr264	21894.8	5.96	21913.0	17.1	21890.5	5.92
pr299	20360.7	31.67	20422.2	44.85	20347.4	28.09
lin318	18521.5	15.96	18596.1	36.9	18511.2	9.70
rd400	5976.3	16.74	6067.8	48.1	5955.0	7.57
fl417	7982.0	0.00	7982.3	0.47	7982.0	0.00
gr431	1033.1	0.25	1037.2	1.64	1033.0	0.25
pr439	51893.5	65.6	52184.0	127.55	51849.7	39.30
pcb442	19796.7	35.51	20079.2	42.39	19729.3	50.90

Table 2.5 Success rates of searching NEN, GEEN, and R2NEN

instance type	$\|V\|$	r	$\|V\|/r$	**NEN**	**GEEN**	**R2NEN**
Grouped Euclidean	125	25	5	54%	49%	72%
	500	100	5	55%	41%	76%
	600	20	30	58%	54%	74%
	1280	64	20	63%	45%	70%
Random Euclidean	250	50	5	74%	30%	95%
	400	20	20	59%	42%	88%
	600	20	30	57%	53%	81%
Non-Euclidean	200	20	10	78%	43%	60%
	500	100	5	80%	16%	68%
	600	20	30	79%	49%	56%
TSBlib-based	n.a.	n.a.	5	55%	44%	67%

there is also one case (ts225) where it was able to obtain a slightly better result than NSB-VNS and COMB-VNS. For three instances, all algorithms obtained the same results, but otherwise NSB-VNS and especially COMB-VNS dominated. COMB-VNS was only beaten on two instances and yielded better results than NSB-VNS and GESB-VNS on nine instances. Again, a Wilcoxon rank test indicates the superiority of COMB-VNS over the other methods with an error probability of less than 1%.

Knowing now that COMB-VNS generally performs best, it is interesting to study the contributions of the individual neighborhood structures. For this purpose, Table 2.5 shows average *success rates* of searching NEN, GEEN, and R2NEN in whole runs of COMB-VNS. By success rate we refer here to the ratio of local searches that

Table 2.6 Average relative gains of searching NEN, GEEN, and R2NEN

instance type	$\|V\|$	r	$\|V\|/r$	NEN	GEEN	R2NEN
Grouped Euclidean	125	25	5	18.02%	63.40%	13.61%
	500	100	5	23.17%	45.19%	28.32%
	600	20	30	13.60%	75.85%	7.29%
	1280	64	20	16.72%	40.08%	20.45%
Random Euclidean	250	50	5	16.90%	48.52%	16.06%
	400	20	20	16.14%	66.75%	15.13%
	600	20	30	21.30%	63.43%	13.65%
Non-Euclidean	200	20	10	10.89%	48.07%	11.92%
	500	100	5	11.74%	60.04%	24.75%
	600	20	30	11.78%	74.80%	10.65%
TSBlib-based	n.a.	n.a.	5	16.58%	60.71%	15.64%

yielded an improved solution divided by the total number of searches of the respective neighborhood structures in percent. In addition, Table 2.6 shows *average relative gains*, i.e., the neighborhood structures' cummulated objective value improvements in relation to the difference between the starting and final solutions' objective values. We can observe that, in general, each neighborhood structure contributes substantially to the whole success. All three neighborhood structures have in almost all cases success rates of over 40%. NEN and R2NEN are more frequently successful though. In contrast, GEEN has—with values of around 60%—significantly higher gains than NEN and R2NEN. Regarding the different instance sets, we can see that the success ratio of GEEN tends to increase with the (average) number of nodes per cluster, which can be explained by there being more node choices available in DP.

While this concludes our discussion of VNS variants for the GMST problem at this point, we will present some further improvements by an additional MIP-based large neighborhood search in Section 4.3 and solution archives avoiding duplicates in Section 6.3.

2.3 Other Applications of Decoder-Based Hybrids

Related to the GMST problem, there are also several other generalized variants of network design and routing problems, where similar incomplete solution representations and dual-representation VNS approaches are, or might be, meaningful. They have in common that a clustered graph is given and a certain structure (subgraph) of minimum cost shall be found that includes exactly one node from each cluster. Examples include the generalized traveling salesman problem [139], generalized vehicle routing problems [22], the generalized minimum edge biconnected network problem [172], and the generalized minimum vertex biconnected network problem [172].

Other examples, where an incomplete solution representation in conjunction with a decoder appears appealing are MIP problems, involving discrete as well as continuous variables. They can be approached by letting a metaheuristic focus on finding good settings for the discrete variables only and augmenting each candidate solution by applying an efficient LP solver as decoder. Such approaches are, for example, described in conjunction with GRASP by Neto and Pedroso [208] and in conjunction with tabu search by Pedroso [217].

Glover [112] suggested an alternative parametric tabu search for heuristically solving MIP problems. A current search point is indirectly represented by the solution to the LP relaxation of the MIP plus additional goal conditions restricting the domains of a subset of the integer variables. Instead of considering the goal conditions directly as hard constraints when applying the LP solver, they are relaxed and brought into the objective function similarly to Lagrangian relaxation. In this way, the approach can also be applied to problems where it is hard to find any feasible integer solution.

In fact, incomplete and/or indirect solution representations in conjunction with "intelligent" decoders can be found quite often in the literature. The main reason for their popularity is that they frequently allow efficient consideration of certain substructures of a given problem, typically also getting rid of complicating constraints in an elegant way. The metaheuristic framework then becomes simpler and frequently more standard neighborhood structures or variation operators can be applied.

Our examples for the GMST are quite specific, especially with the additional feature of combining two complementary solution representations with their respective neighborhood structures. In the NSB as well as the GESB representations, both decoders are *exact* in the sense that they are guaranteed to always yield an optimally augmented solution, i.e., if the "right" encoded/incomplete solution is given, we can be sure to obtain a globally optimal solution. Similarly, this also holds for the above sketched approach for solving MIP problems.

This is, however, not necessarily always the case. Frequently, decoders are heuristics themselves, and their limitations might also imply that certain, possibly promising, solutions cannot be reached at all. Nevertheless, such approaches might be highly useful for complex practical problems.

A prominent, quite generic way of indirectly representing solutions is by means of *permutations* of solution elements. The decoder is then often a constructive heuristic which composes a solution by adding/considering the solution elements in the order specified by the permutation. For example, such approaches are frequently used for scheduling or packing problems [154]. The solution elements are in these cases jobs or events to be scheduled or items to be packed, respectively. The metaheuristic must then provide neighborhood structures or variation operators suitable for permutations.

Another related indirect solution representation that makes fundamental use of decoders are *random key* approaches, originally proposed by Bean in the context of genetic algorithms [14] and more recently applied to a larger number of problems by Resende et al. [121]. A solution is represented by a vector of real values associated again with the solution elements. For initial solutions, these values are typically independently set to random values of a certain interval, e.g. $[0,1)$ (therefore the name "random keys"). The decoder sorts all solution elements according to their random keys and then performs as in a permutation-based method in a problem-specific way. The main advantage of this approach is that the metaheuristic's search space is even more basic (i.e., $[0,1)^n$ for n solution elements) and standard variation operators like uniform crossover and mutation by setting randomly selected keys to new random values can be used, entirely independent of the targeted problem.

One more related indirect representation technique that also makes use of a vector of real numbers are *weight-codings*, originally proposed in the context of genetic algorithms for constrained MST problems [213] but afterwards also applied to a variety of other combinatorial optimization problems. The main difference from random keys is that in weight-codings the real number vectors are used to *distort* the problem instance, e.g., by adding the values to a graph's edge costs, in order to obtain different solutions from a deterministic greedy heuristic.

Obviously, the performance of a decoder-based approach strongly depends on the quality and speed of the decoder. At this point, however, we also need to give a word of warning concerning highly indirect representations: While they appear appealing for many problems due to their generality and frequently easy applicability, one should also carefully consider the relationship of a problem's original solution space and the metaheuristic's search space and operators. Typically a metaheuristic can only be effective if the solutions in a searched neighborhood or produced by the respective variation operators are most of the time indeed strongly related. Representations and decoders that are too artificial might yield a distorted search space with many new local optima introduced or similar solutions becoming strongly separated. The heuristic search might then not be effective anymore. In fact, there is typically a tradeoff between generality and performance of metaheuristic approaches.

Last but not least, in this context we also want to refer to *hyper-heuristics* [44], which try to achieve a high level of generality, i.e., applicability to a wide range of problems, but at the same time allow a large degree of problem-specialization. Hyper-heuristics act on an indirect search space defined by a pool of lower-level heuristics or heuristic components targeted to the actual problem. Lower-level heuristics can range from simple greedy construction steps, local improvement strategies to advanced algorithmic techniques such as even other metaheuristics. Roughly speaking, hyper-heuristics can be seen as a high-level methodology which, when a particular problem instance or class of instances is given, try to automatically produce an adequate combination of the provided algorithmic components to effectively solve the given problem. Successful applications of hyper-heuristics can in particular be found in the domains of scheduling and planning, packing, and constraint satisfaction. A recent state-of-the-art review can be found in [45].

Chapter 3
Hybridization Based on Problem Instance Reduction

This chapter presents an example of a hybrid metaheuristic for optimization based on the following general idea. General MIP solvers such as CPLEX and GUROBI are often very effective up to a certain, problem-specific instance size. When given a problem instance too large to be directly solved by a MIP solver, it might be possible to reduce the problem instance in a clever way such that the resulting reduced problem instance contains high-quality solutions—or even optimal solutions—to the original problem instance and such that the reduced problem instance can be effectively solved by the MIP solver. In this way, it would be possible to take profit from valuable Operations Research expertise that went into the development of the MIP solvers, even in the context of problem instances too large to be solved directly.

The outline of this chapter is as follows. After a more detailed description of the general idea sketched above, a problem-independent algorithm for optimization labeled CONSTRUCT, MERGE, SOLVE & ADAPT will be described, which makes use of this general idea. This algorithm works roughly as follows. First, a reduced sub-instance of the original problem instance is generated by means of merging the solution components found in probabilistically generated solutions to the original problem instance; second, an exact solver is applied to the reduced sub-instance in order to obtain a promising solution to the original problem instance; and, third, the sub-instance is adapted for the next iteration of the algorithm. The *Minimum Common String Partition* (MCSP) problem is chosen as a test case in order to demonstrate the application of the algorithm. Obtained results show that the proposed algorithm is superior to its pure components: (1) the iterative probabilistic construction of solutions, and (2) the application of an ILP solver to the original problem instance.

3.1 General Idea

The algorithm presented in this chapter is based on the following general idea. Imagine it were possible to identify a substantially reduced sub-instance of a given problem instance such that the sub-instance contains high-quality solutions to the original problem instance. This might allow one to apply an exact technique—such as, for example, an ILP solver—with reasonable computational effort to the reduced sub-instance in order to obtain a high-quality solution to the original problem instance. This is for the following reason. For many combinatorial optimization problems the field of mathematical programming—and integer linear programming in particular—provides powerful tools; see Section 1.5 and [299] for a more comprehensive introduction. MIP solvers are, in general, based on a tree search framework but further consider the solutions of linear programming relaxations for the problem at hand (besides primal heuristics) in order to obtain lower and upper bounds. To tighten these bounds, various kinds of additional inequalities are typically dynamically identified and added as cutting planes to the ILP model, yielding a branch-and-cut algorithm. Frequently, such ILP approaches are highly effective for small to medium sized instances of hard problems, even though they often do not scale well enough to large instances relevant in practice. Therefore, when a problem instance can be sufficiently reduced, a MIP solver might be very effective in solving the reduced problem instance.

For the following discussion remember that, given a problem instance $\mathcal{I}$ to a generic problem $\mathcal{P}$, set C represents the set of all possible components of which solutions to the problem instance are composed. C is called the complete set of solution components with respect to $\mathcal{I}$. Note that, given an integer linear (or non-linear) programming model for problem $\mathcal{P}$, a generic way of defining the set of solution components is to say that each combination of a variable with one of its values is a solution component. Moreover, in the context of this chapter a valid solution S to $\mathcal{I}$ is represented as a subset of the solution components C, that is, $S \subseteq C$. Finally, set $C' \subseteq C$ contains the solution components that belong to a restricted problem instance, that is, a sub-instance of $\mathcal{I}$. For simplicity reasons, C' will henceforth be called a sub-instance. Imagine, for example, the input graph in the case of the traveling salesman problem. The set of all edges can be regarded as the set of all possible solution components C. Moreover, the edges belonging to a tour S—that is, a valid solution—form the set of solution components that are contained in S.

3.1.1 The Construct, Merge, Solve & Adapt Algorithm

The CONSTRUCT, MERGE, SOLVE & ADAPT (CMSA) algorithm works roughly as follows. At each iteration, the algorithm deals with the incumbent sub-instance C'. Initially this sub-instance is empty. The first step of each iteration consists in *generating* a number of feasible solutions to the original problem instance $\mathcal{I}$ in a probabilistic way. In the second step, the solution components involved in these solutions are added to C' and an exact solver is applied in order to *solve* C' to optimality, if given enough computation time. Otherwise, the best solution found is returned by the exact solver. The third step consists in *adapting* sub-instance C' by

Algorithm 17 CONSTRUCT, MERGE, SOLVE & ADAPT (CMSA)

```
 1: given: problem instance 𝒥, values for parameters n_a and age_max
 2: S_bsf ← NULL; C' ← ∅
 3: age[c] ← 0 for all c ∈ C
 4: while CPU time limit not reached do
 5:     for i ← 1,...,n_a do
 6:         S ← ProbabilisticSolutionGeneration(C)
 7:         for all c ∈ S and c ∉ C' do
 8:             age[c] ← 0
 9:             C' ← C' ∪ {c}
10:         end for
11:     end for
12:     S'_opt ← ApplyExactSolver(C')
13:     if S'_opt is better than S_bsf then S_bsf ← S'_opt
14:     Adapt(C', S'_opt, age_max)
15: end while
16: return s_bsf
```

removing some of the solution components guided by an aging mechanism. In other words, the CMSA algorithm is in principle applicable to any problem for which (1) a way of (probabilistically) generating solutions can be found and (2) a strategy for solving smaller instances of the problem to optimality is known.

In the following we describe the CMSA algorithm, which is shown in pseudo-code in Algorithm 17, in more detail. The main loop of the algorithm is executed while the CPU time limit is not reached. It consists of the following actions. First, the best-so-far solution S_{bsf} is initialized to NULL, indicating that no such solution exists yet, and the restricted problem instance C' to the empty set. Then, at each iteration a number of n_a solutions is probabilistically generated, see function ProbabilisticSolutionGeneration(C) in line 6 of Algorithm 17. The components of all these solutions are added to set C'. The age of a newly added component c—denoted here by $age[c]$—is set to 0. After the construction of n_a solutions, an exact solver is applied to find a possibly optimal solution S'_{opt} to the restricted problem instance C', see function ApplyExactSolver(C') in line 12 of Algorithm 17. If S'_{opt} is better than the current best-so-far solution S_{bsf}, solution S'_{opt} is adopted as the new best-so-far solution (line 13). Next, sub-instance C' is adapted, based on solution S'_{opt} and on the age values of the solution components. This is done in function Adapt(C', S'_{opt}, age_{max}) in line 14 as follows. First, the age of each solution component in $C' \setminus S'_{opt}$ is incremented while the age of each solution component in $S'_{opt} \subseteq C'$ is re-initialized to zero. Then, those solution components from C' whose age has reached the maximum component age (age_{max}) are deleted from C'. The motivation behind the aging mechanism is that components which never appear in the solutions of C' returned by the exact solver should be removed from C' after some time, because they would otherwise slow down the exact solver on the long term. In contrast, components which appear in the solutions returned by the exact solver seem to be useful and should therefore remain in C'.

In summary, the behavior of the general CMSA algorithm depends substantially on the values of two parameters: the number of solution constructions per iteration (n_a) and the maximum allowed age (age_{max}) of solution components. This completes the general description of the algorithm.

3.2 Application to Minimum Common String Partition

The *Minimum Common String Partition* (MCSP) problem [54] is used as a test case for the CMSA algorithm. This problem is an NP-hard combinatorial optimization problem from the bioinformatics field, which can technically be described as follows. Given are two input strings s_1 and s_2 of length n over a finite alphabet Σ. The two strings are *related*, which means that each letter appears the same number of times in each of them. Note that this definition implies that s_1 and s_2 have the same length (n). A valid solution to the MCSP problem is obtained by partitioning s_1 into a set P_1 of non-overlapping substrings, and s_2 into a set P_2 of non-overlapping substrings, such that $P_1 = P_2$. Moreover, the goal is to find a valid solution such that $|P_1| = |P_2|$ is minimal. Consider the following example. Given are DNA sequences $s_1 = \mathbf{AGACTG}$ and $s_2 = \mathbf{ACTAGG}$. As **A** and **G** appear twice in both input strings, and **C** and **T** appear once, the two strings are related. A trivial valid solution can be obtained by partitioning both strings into substrings of length one, that is, $P_1 = P_2 = \{\mathbf{A}, \mathbf{A}, \mathbf{C}, \mathbf{T}, \mathbf{G}, \mathbf{G}\}$. The objective function value of this solution is six. However, the optimal solution, with objective function value three, is $P_1 = P_2 = \{\mathbf{ACT}, \mathbf{AG}, \mathbf{G}\}$.

The MCSP problem was introduced by Chen et al. [54] due to its relation to genome rearrangement. More specifically, it has applications in biological questions such as: May a given DNA string possibly be obtained by rearrangements of another DNA string? The general problem has been shown to be NP-hard even in very restrictive cases [118]. The MCSP has been considered quite extensively by researchers dealing with the approximability of the problem. Cormode and Muthukrishnan [66], for example, proposed an $O(\log n \log^* n)$-approximation for the *edit distance with moves* problem, which is a more general case of the MCSP problem. Shapira and Storer [267] expanded on this result. Other approximation approaches for the MCSP problem were proposed in [162]. In this context, Chrobak et al. [56] studied a simple greedy approach for the MCSP problem, showing that the approximation ratio concerning the 2-MCSP problem, in which each letter of the alphabet appears at most twice, is three, and for the 4-MCSP problem the approximation ratio is $\Omega(\log(n))$. In the case of the general MCSP problem, the approximation ratio lies between $\Omega(n^{0.43})$ and $O(n^{0.67})$, assuming that the input strings use an alphabet of size $O(\log(n))$. Later Kaplan and Shafir [150] raised the lower bound to $\Omega(n^{0.46})$. Kolman proposed a modified version of the simple greedy algorithm with an approximation ratio of $O(k^2)$ for the k-MCSP [161]. Recently, Goldstein and Lewenstein proposed a greedy algorithm for the MCSP problem that runs in $O(n)$ time [119].

He [131] introduced a greedy algorithm with the aim of obtaining better average results. To our knowledge, the only metaheuristic approaches that have been proposed in the literature for the MCSP problem are (1) the MAX-MIN Ant System by Ferdous and Sohel Rahman [93, 94] and (2) the probabilistic tree search algorithm by Blum et al. [36]. In both works the proposed algorithm is applied to a range of artificial and real DNA instances from [93]. Finally, as mentioned before, the first ILP model for the MCSP problem, together with an ILP-based heuristic, was described in [37].

The remainder of this section describes the application of the CMSA algorithm presented in the previous section to the MCSP problem. For this purpose we define the set of solution components and the structure of valid (partial) solutions as follows. Henceforth, a *common block* c_i of input strings s_1 and s_2 is denoted as a triple (t_i, k_i^1, k_i^2) where t_i is a string that can be found starting at position $1 \leq k_i^1 \leq n$ in string s_1 and starting at position $1 \leq k_i^2 \leq n$ in string s_2. Moreover, let $C = \{c_1, \ldots, c_m\}$ be the (ordered) set of all possible common blocks of s_1 and s_2.[1] Note that C is the set of all solution components. Given the definition of C, any valid solution S to the MCSP problem can be expressed as a subset of C—that is, $S \subset C$—such that:

1. $\sum_{c_i \in S} |t_i| = n$, that is, the sum of the length of the strings corresponding to the common blocks in S is equal to the length of the input strings.
2. For any two common blocks c_i, $c_j \in S$ it holds that their corresponding strings neither overlap in s_1 nor in s_2.

Moreover, any (valid) partial solution S_{partial} is a subset of C fulfilling the following conditions: (1) $\sum_{c_i \in S_{\text{partial}}} |t_i| < n$ and (2) for any two common blocks $c_i, c_j \in S_{\text{partial}}$ it holds that their corresponding strings neither overlap in s_1 nor in s_2. Note that any valid partial solution can be extended to be a valid solution. Furthermore, given a partial solution S_{partial}, set $Ext(S_{\text{partial}}) \subset C$ denotes the set of common blocks that may be used in order to extend S_{partial} such that the result is again a valid (partial) solution.

3.2.1 Probabilistic Solution Generation

Next we describe the implementation of function ProbabilisticSolutionGeneration(C) in line 6 of Algorithm 17. The construction of a solution (see Algorithm 18) starts with the empty partial solution S_{partial}, that is, $S_{\text{partial}} \leftarrow \emptyset$. At each construction step, a solution component c^* from $Ext(S_{\text{partial}})$ is chosen and added to S_{partial}. This is done until S_{partial} is a complete solution. The choice of c^* is done as follows. First, a value $\delta \in [0, 1)$ is chosen uniformly at random. In case $\delta \leq d_{\text{rate}}$, c^* is chosen such that $|t_{c^*}| \geq |t_c|$ for all $c \in Ext(S_{\text{partial}})$, that is, one of the common blocks whose substring is of maximal size is chosen. Otherwise, a candidate list L containing the (at most) l_{size} longest common blocks from $Ext(S_{\text{partial}})$ is built, and b^* is chosen from

[1] The way in which C is ordered is of no importance.

Algorithm 18 Probabilistic Solution Generation

1: **given:** $s_1, s_2, d_{\text{rate}}, l_{\text{size}}$
2: $S_{\text{partial}} \leftarrow \emptyset$
3: **while** S_{partial} is not a complete solution **do**
4: choose a random number $\delta \in [0,1]$
5: **if** $\delta \leq d_{\text{rate}}$ **then**
6: choose c^* such that $|t_{c^*}| \geq |t_c|$ for all $c \in Ext(S_{\text{partial}})$
7: $S_{\text{partial}} \leftarrow S_{\text{partial}} \cup \{c^*\}$
8: **else**
9: let $L \subseteq Ext(S_{\text{partial}})$ contain the (at most) l_{size} longest common blocks from $Ext(S_{\text{partial}})$
10: choose randomly c^* from L
11: $S_{\text{partial}} \leftarrow S_{\text{partial}} \cup \{c^*\}$
12: **end if**
13: **end while**
14: **return** complete solution $S \leftarrow S_{\text{partial}}$

L uniformly at random. In other words, the greediness of this procedure depends on the pre-determined values of d_{rate} (determinism rate) and l_{size} (candidate list size). Both are input parameters of the algorithm.

3.2.2 Solving Reduced Sub-instances

The last component of Algorithm 17 which remains to be described is the implementation of function ApplyExactSolver(C') in line 12. In the case of the MCSP problem we make use of the ILP model proposed in [37] and the MIP solver CPLEX for solving it. The model for the complete set C of solution components can be described as follows. First, two binary $m \times n$ matrices M^1 and M^2 are defined. In both matrices, row $1 \leq i \leq m$ corresponds to common block $c_i \in C$. Moreover, a column $1 \leq j \leq n$ corresponds to position j in input string s_1, respectively s_2. In general, the entries of matrix M^1 are set to zero. However, in each row i, the positions that string t_i (of common block c_i) occupies in input string s_1 are set to one. Correspondingly, the entries of matrix M^2 are set to zero, apart from the fact that in each row i, the positions occupied by string t_i in input string s_2 are set to one. Henceforth, the position (i,j) of a matrix M^k, $k \in \{1,2\}$, is denoted by $M^k_{i,j}$. Finally, we introduce for each common block $c_i \in C$ a binary variable x_i. With these definitions the MCSP can be expressed in the form of the following ILP model.

$$\text{minimize} \quad \sum_{i=1}^{m} x_i \tag{3.1}$$

$$\text{subject to} \quad \sum_{i=1}^{m} M^1_{i,j} \cdot x_i = 1 \qquad \forall j = 1,\ldots,n \tag{3.2}$$

$$\sum_{i=1}^{m} M^2_{i,j} \cdot x_i = 1 \qquad \forall j = 1,\ldots,n \tag{3.3}$$

$$x_i \in \{0,1\} \qquad \forall i = 1,\ldots,m \tag{3.4}$$

The objective function minimizes the number of selected common blocks. Equations (3.2) ensure that the strings corresponding to the selected common blocks do not overlap in input string s_1, while equations (3.3) make sure that the strings corresponding to the selected common blocks do not overlap in input string s_2. The condition that the length of the strings corresponding to the selected common blocks is equal to n is implicitly obtained from these two sets of constraints.

As an example, consider the small problem instance that was mentioned in the course of the explanation of the MCSP problem. The complete set of common blocks (C) as induced by input strings $s_1 = \mathbf{AGACTG}$ and $s_2 = \mathbf{ACTAGG}$ is

$$C = \left\{ \begin{array}{l} c_1 = (\mathbf{ACT}, 3, 1) \\ c_2 = (\mathbf{AG}, 1, 4) \\ c_3 = (\mathbf{AC}, 3, 1) \\ c_4 = (\mathbf{CT}, 4, 2) \\ c_5 = (\mathbf{A}, 1, 1) \\ c_6 = (\mathbf{A}, 1, 4) \\ c_7 = (\mathbf{A}, 3, 1) \\ c_8 = (\mathbf{A}, 3, 4) \\ c_9 = (\mathbf{C}, 4, 2) \\ c_{10} = (\mathbf{T}, 5, 3) \\ c_{11} = (\mathbf{G}, 2, 5) \\ c_{12} = (\mathbf{G}, 2, 6) \\ c_{13} = (\mathbf{G}, 6, 5) \\ c_{14} = (\mathbf{G}, 6, 6) \end{array} \right\}$$

Given set C, matrices M^1 and M^2 are as follows:

$$M^1 = \begin{pmatrix} 0&0&1&1&1&0 \\ 1&1&0&0&0&0 \\ 0&0&1&1&0&0 \\ 0&0&0&1&1&0 \\ 1&0&0&0&0&0 \\ 1&0&0&0&0&0 \\ 0&0&1&0&0&0 \\ 0&0&1&0&0&0 \\ 0&0&0&1&0&0 \\ 0&0&0&0&1&0 \\ 0&1&0&0&0&0 \\ 0&1&0&0&0&0 \\ 0&0&0&0&0&1 \\ 0&0&0&0&0&1 \end{pmatrix}, \quad M^2 = \begin{pmatrix} 1&1&1&0&0&0 \\ 0&0&0&1&1&0 \\ 1&1&0&0&0&0 \\ 0&1&1&0&0&0 \\ 1&0&0&0&0&0 \\ 0&0&0&1&0&0 \\ 1&0&0&0&0&0 \\ 0&0&0&1&0&0 \\ 0&1&0&0&0&0 \\ 0&0&1&0&0&0 \\ 0&0&0&0&1&0 \\ 0&0&0&0&0&1 \\ 0&0&0&0&1&0 \\ 0&0&0&0&0&1 \end{pmatrix}$$

The optimal solution to this instance is $S^* = \{c_1, c_2, c_{14}\}$. It can easily be verified that this solution respects equations (2) and (3) of the ILP model.

Note that this ILP model can also be solved for any subset C' of C. This is achieved by replacing all occurrences of C with C', and by replacing m with $|C'|$. The solution of such an ILP corresponds to a feasible solution to the original problem instance as long as C' contains at least one feasible solution to the original problem instance. Due to the way in which C' is generated (see Section 3.2.1), this condition is fulfilled.

3.2.3 Experimental Evaluation

The described application of CMSA to the MCSP problem was implemented in ANSI C++ using GCC 4.7.3 for compiling the software. Moreover, both the complete ILP models and the reduced ILP models within CMSA were solved with IBM ILOG CPLEX 12.1. The experimental evaluation was conducted on a cluster of 32 PCs with Intel Xeon CPU X5660 CPUs of 2 nuclei of 2.8 GHz and 48 GB of RAM. The following algorithms were considered for the comparison:

1. GREEDY: the greedy approach from [131].
2. SOLCONST: the CMSA algorithm without the execution of the MIP solver in line 12 of Algorithm 17. This is done in order to determine the contribution of the MIP solver to the solution quality obtained by the full CMSA algorithm.
3. $\text{ILP}_{\text{compl}}$: the application of CPLEX to the complete ILP for each considered problem instance.
4. CMSA: the full CMSA approach.

For the comparison of these algorithms we generated a set of 600 benchmark instances. In more detail, this benchmark set consists of 20 randomly generated instances for each combination of $n \in \{200, 400, \ldots, 1800, 2000\}$, where n is the length of the input strings, and alphabet size $|\Sigma| \in \{4, 12, 20\}$. Ten of these instances were generated with an equal probability for each letter of the alphabet. The probability for each letter $l \in \Sigma$ to appear at a certain position of the input strings is $1/|\Sigma|$. The resulting set of 300 benchmark instances of this type is labeled LINEAR. The other 10 instances per combination of n and $|\Sigma|$ are generated with a probability for each letter $l \in \Sigma$ to appear at a certain position of the input strings of $ord(l)/\sum_{i=1}^{|\Sigma|} i$, where $ord(l)$ refers to the position of l in the ordered alphabet set. The resulting set of 300 benchmark instances of this second type is labeled SKEWED.

3.2.3.1 Tuning of CMSA and SOLCONST

CMSA has several parameters for which well-working values must be found: (1) the number n_a of solution constructions per iteration, (2) the maximum allowed age $age_{\max}$ of solution components, (3) the determinism rate d_{rate}, (4) the candidate list size l_{size}, and (5) the maximum time t_{max} (in seconds) allowed for CPLEX per application to a sub-instance. The last parameter is necessary, because even when applied to reduced problem instances, CPLEX might still need too much computation time for solving such sub-instances to optimality. In any case, CPLEX always returns the best feasible solution found within the given computation time.

The automatic configuration tool irace [179] was used for the fine-tuning of the five parameters. In fact, irace was applied to tune CMSA separately for each instance size in $\{200, 400, \ldots, 1800, 2000\}$. In total, 12 *training instances* for each of the 10 different instance sizes were generated for tuning: two instances of type LINEAR and two instances of type SKEWED for each alphabet size from $\{4, 12, 20\}$. The tuning process for each instance size was given a budget of 1000 runs of CMSA, where each run was given a computation time limit of 3600 CPU seconds. Finally, the following domains were chosen for the five parameters of CMSA:

- $n_a \in \{10, 30, 50\}$
- $age_{max} \in \{1, 5, 10, \inf\}$, with inf referring to the case in which components are never removed from C'
- $d_{rate} \in \{0.0, 0.5, 0.9\}$; in case of 0.0 the selection of solution component c^* (line 6 of Algorithm 18) is always done randomly from the candidate list, while with 0.9 the solution constructions are nearly deterministic
- $l_{size} \in \{3, 5, 10\}$
- $t_{max} \in \{60, 120, 240, 480\}$ seconds

The 10 applications of irace produced the 10 configurations for CMSA as shown in Table 3.2a. The following tendencies can be observed. First of all, with growing instance size, more time should be given to individual applications of CPLEX to sub-instances of the original problem instance. Second, irrespective of the instance size, candidate list sizes (l_{size}) smaller than five seem to be too restrictive. Third, again irrespective of the instance size, less than 30 solution constructions per iteration seem to be insufficient. Presumably, when only a few solution constructions per iteration are performed, the resulting change in the corresponding sub-instances is not large enough and, therefore, some applications of CPLEX result in wasted computation time. Finally, considering the obtained values of d_{rate} for instance sizes from 200 to 1600, the impression is that with growing instance size the degree of greediness in the solution construction should grow. However, the settings of d_{rate} for $n \in \{1800, 2000\}$ seem to contradict this assumption.

In addition to tuning experiments for CMSA, we also performed tuning experiments for SOLCONST. The parameters involved in the construction of solutions are d_{rate} and l_{size}. For the fine-tuning of these parameters in the context of SOLCONST we used the same training instances, the same budget of 1000 runs, and the same parameter value ranges as for the tuning of CMSA. The obtained parameter values per instance size are displayed in Table 3.2b. Interestingly, the values obtained for parameter d_{rate} show a tendency which is contrary to the values of d_{rate} in CMSA, that is, with growing instance size the degree of greediness in the solution construction should decrease.

Table 3.1 Parameter settings produced by irace for the 10 different instance sizes

n	n_a	age_{max}	d_{rate}	l_{size}	t_{max}
200	50	*inf*	0.0	10	60
400	50	10	0.0	10	60
600	50	10	0.5	10	60
800	50	10	0.5	10	240
1000	50	10	0.9	10	480
1200	50	10	0.9	10	480
1400	50	*inf*	0.9	5	480
1600	50	5	0.9	10	480
1800	30	10	0.5	5	480
2000	50	10	0.0	10	480

(a) Tuning results for CMSA

n	d_{rate}	l_{size}
200	0.9	5
400	0.9	3
600	0.9	10
800	0.5	5
1000	0.5	5
1200	0.5	3
1400	0.5	5
1600	0.0	5
1800	0.0	10
2000	0.0	10

(b) Tuning results for SOLCONST

3.2.3.2 Results

As described before, the benchmark set consists of 300 instances of type LINEAR and another 300 instances of type SKEWED. The results for instances of type LINEAR are given in the three Tables 3.3a–3.3c, in terms of one table per alphabet size, and those for instances of type SKEWED are given in the three Tables 3.4a–3.4c. Average values over 10 random instances of the same characteristics are listed. Each algorithm included in the comparison was applied exactly once to each problem instance.

The structure of all tables is as follows. The first column provides the instance size. The second column contains the final objective values of GREEDY. The third column contains the final objective values obtained by SOLCONST. The next three table columns are dedicated to the presentation of the results the complete ILP model $\text{ILP}_{\text{compl}}$ yielded. The first one of these columns provides the values of the best solutions found within 3600 CPU seconds, the second column the computation time in seconds. In cases of having solved the corresponding problem to optimality, this column only displays one value indicating the time needed by CPLEX to solve the problem. Otherwise, this column provides two values in the form X/Y, where X corresponds to the time at which CPLEX was able to find the first valid solution, and Y to the time at which CPLEX found the best solution within 3600 CPU seconds. Finally, the third one of the columns dedicated to $\text{ILP}_{\text{compl}}$ shows the optimality gap, which refers to the relative difference between the value of the best valid solution and the current lower bound at the time of stopping a run. The last two columns of each table are dedicated to the presentation of the results obtained by CMSA. The first column provides the values of the solutions found by CMSA within 3600 CPU seconds. The last column indicates the average computation time needed by CMSA to find the best solution of each run. The best result for each row is marked by a gray background and the last row of each table provides averages over the whole table.

Table 3.2 Results for the instances of set LINEAR

n	GREEDY	SOLCONST	ILP$_{compl}$			CMSA	
	mean	**mean**	**mean**	**time**	**gap**	**mean**	**time**
200	75.0	68.7	63.5	4/104	0.0%	63.7	608
400	133.4	126.1	115.7	108/2081	6.8%	116.4	1381
600	183.7	177.5	162.2	513/1789	9.4%	162.9	1918
800	241.1	232.7	246.8	1671/1671	23.8%	212.4	2434
1000	287.0	280.4	n/a	n/a	n/a	256.9	2623
1200	333.8	330.4	n/a	n/a	n/a	303.3	2369
1400	385.5	378.9	n/a	n/a	n/a	351.0	2452
1600	432.3	427.1	n/a	n/a	n/a	400.6	1973
1800	477.4	474.2	n/a	n/a	n/a	445.4	2194
2000	521.6	520.7	n/a	n/a	n/a	494.0	1744
avg.	307.1	301.7	n/a	n/a	n/a	280.7	1969

(a) Results for instances with $\Sigma = 4$

n	GREEDY	SOLCONST	ILP$_{compl}$			CMSA	
	mean	**mean**	**mean**	**time**	**gap**	**mean**	**time**
200	127.3	122.1	119.2	1/1	0.0	119.2	22
400	228.9	223.5	208.9	7/51	0.0	209.4	892
600	322.2	318.7	291	47/1277	0.9	293.8	1433
800	411.4	408.1	368.7	147/2405	1.6	373.2	1484
1000	499.2	494.9	453.4	395/2084	3.8	449.9	2651
1200	586.0	585.6	536.6	784/3188	4.7	531.0	2318
1400	666.0	664.6	684.1	1667/1667	15.8	606.9	2467
1600	754.4	754.6	773.5	2648/2648	16.0	694.8	2392
1800	827.3	833.0	n/a	n/a	n/a	773.6	1484
2000	913.5	916.2	n/a	n/a	n/a	849.6	2967
avg.	533.6	532.1	n/a	n/a	n/a	490.1	1811

(b) Results for instances with $\Sigma = 12$

n	GREEDY	SOLCONST	ILP$_{compl}$			CMSA	
	mean	**mean**	**mean**	**time**	**gap**	**mean**	**time**
200	149.2	146.6	146.2	1/1	0.0%	146.2	2
400	274.5	268.8	261.5	2/2	0.0%	261.9	80
600	389.2	383.5	362.3	10/15	0.0%	366.6	364
800	495.8	492.3	456.1	43/700	0.0%	463.1	804
1000	600.6	597.5	547.1	125/1737	0.6%	555.0	529
1200	706.1	707.8	642.2	296/2732	1.3%	648.5	1372
1400	801.1	804.0	737.9	559/2314	3.1%	737.7	2334
1600	899.8	903.1	861.3	966/2885	6.6%	825.7	2251
1800	996.8	1000.1	1012.9	1559/1845	12.6%	917.6	2437
2000	1097.8	1102.6	1136.0	2349/2349	14.4%	1024.9	2924
avg.	641.1	640.6	616.35	591/1458	3.9%	594.4	1310

(c) Results for instances with $\Sigma = 20$

Table 3.3 Results for the instances of set SKEWED

n	GREEDY	SOLCONST	ILP$_{\text{compl}}$			CMSA	
	mean	**mean**	**mean**	**time**	**gap**	**mean**	**time**
200	68.7	62.8	57.4	10/217	0.0%	57.5	903
400	120.3	115.0	105.3	168/1330	7.6%	105.1	1314
600	170.6	163.8	149.7	938/2193	10.1%	150.4	1500
800	219.8	213.3	224	2600/2600	22.9%	196.5	2303
1000	268.6	261.7	n/a	n/a	n/a	240.2	2692
1200	313.8	309.0	n/a	n/a	n/a	285	2785
1400	358.7	352.2	n/a	n/a	n/a	327.6	2888
1600	400.9	397.9	n/a	n/a	n/a	376.0	2171
1800	440.6	442.1	n/a	n/a	n/a	417.7	2162
2000	485.0	481.2	n/a	n/a	n/a	470.2	1222
avg.	284.7	279.9	n/a	n/a	n/a	262.6	1994

(a) Results for instances with $\Sigma = 4$

n	GREEDY	SOLCONST	ILP$_{\text{compl}}$			CMSA	
	mean	**mean**	**mean**	**time**	**gap**	**mean**	**time**
200	117.9	112.7	108.5	1/1	0.0%	108.6	12
400	216.1	208.5	193.4	10/136	0.0%	194.3	1002
600	304.8	301.7	274.5	71/1081	1.2%	277.2	1711
800	389.3	385.4	347.0	248/2725	2.3%	351.0	2177
1000	471.6	468.9	429.4	650/2582	4.9%	424.4	2648
1200	551.1	549.9	559.4	1351/1804	14.9%	500.1	2597
1400	625.7	626.3	645.1	2693/2693	16.7%	570.0	2962
1600	705.6	706.4	n/a	n/a	n/a	643.8	2434
1800	788.4	788.9	n/a	n/a	n/a	723.3	2329
2000	857.8	858.0	n/a	n/a	n/a	797.3	2805
avg.	502.8	500.7	n/a	n/a	n/a	459.0	2068

(b) Results for instances with $\Sigma = 12$

n	GREEDY	SOLCONST	ILP$_{\text{compl}}$			CMSA	
	mean	**mean**	**mean**	**time**	**gap**	**mean**	**time**
200	140.4	135.9	134.7	1/1	0.0%	134.7	8
400	255.5	251.3	240.3	3/4	0.0%	240.6	1080
600	366.8	361.2	336.1	19/101	0.0%	341.1	764
800	466.3	462.7	424.4	80/1119	0.2%	429.8	753
1000	567.6	566.6	514.7	202/2253	0.9%	520.9	1121
1200	661.8	662.4	604.2	469/2064	2.1%	605.7	1869
1400	762.3	760.7	694.4	719/2511	2.8%	693.2	1743
1600	851.2	855.2	863.3	1378/1828	12.3%	780.4	2681
1800	948.7	948.8	969.8	1774/1976	13.9%	870.2	2815
2000	1034.3	1037.7	1061.6	2589/2589	14.4%	967.1	2978
avg.	605.5	604.3	584.4	723/1844	4.7%	558.4	1581

(c) Results for instances with $\Sigma = 20$

The analysis of the results leads to the following conclusions:

- Surprisingly, hardly any differences can be observed in the relative performance of the algorithms for instances of type LINEAR in comparison to instances of type SKEWED. Therefore, all following observations hold both for LINEAR and SKEWED instances.
- Concerning the application of CPLEX to $\text{ILP}_{\text{compl}}$, the alphabet size has a strong influence on the problem difficulty. For instances with $|\Sigma| = 4$, CPLEX is only able to provide feasible solutions within 3600 CPU seconds for input strings of lengths up to 800, both in the context of instances LINEAR and SKEWED. When $|\Sigma| = 12$, CPLEX can provide feasible solutions for input strings of lengths up to 1600 (LINEAR), respectively 1400 (SKEWED). However, starting from $n = 1000$ CPLEX is not competitive with CMSA anymore. Finally, even though CPLEX can provide feasible solutions for all instance sizes concerning the instances with $|\Sigma| = 20$, starting from $n = 1400$ the results are inferior to those of CMSA.
- For rather small instances the differences between the results of CMSA and those of applying CPLEX to $\text{ILP}_{\text{compl}}$ are, again, not statistically significant.
- Remarkably, CMSA outperforms SOLCONST in all cases. This means that the application of the MIP solver within CMSA is essential for its success. Moreover, even though the quality of the probabilistically constructed solutions is rather low, the sub-instances that are obtained by merging these solutions contain high-quality solutions.

In summary, we can state that CMSA is competitive with the application of CPLEX to the original ILP model when the size of the input instances is rather small. The larger the size of the input instances, and the smaller the alphabet size, the greater is the advantage of CMSA over the other algorithms. The validity of these statements can be conveniently observed in Figures 3.1 and 3.2.

Finally, we also provide information about the sizes of the original problem instances (in terms of the number of solution components) in comparison to the average sizes of the sub-instances generated by CMSA. On the one hand, the six solid lines in Figure 3.3 show the evolution of the original instance sizes. These lines are labeled "ILP Σ X Y" where X is the alphabet size and $Y \in \{\text{L}, \text{S}\}$. Hereby, L stands for instances LINEAR and S for instances SKEWED. All six curves show an exponential growth with increasing length of the input strings (n). On the other hand, the six dashed lines show the average size of the sub-instances solved by CPLEX within CMSA. It can be observed that the growth of the sub-instance sizes is rather linear with increasing n. Moreover, the size of these sub-instances is significantly smaller than the size of the original problem instances. This is why CPLEX can solve these sub-instances to (near) optimality in relatively little time.

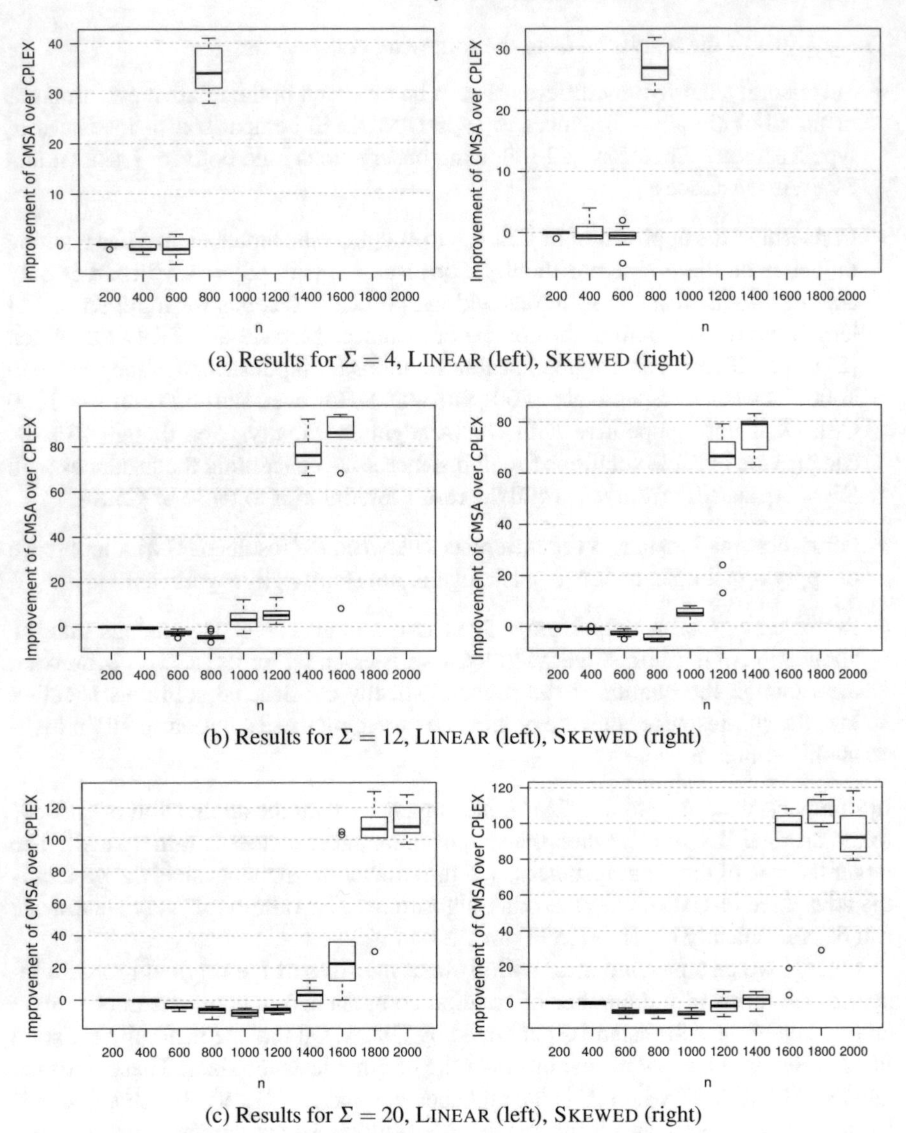

(a) Results for $\Sigma = 4$, LINEAR (left), SKEWED (right)

(b) Results for $\Sigma = 12$, LINEAR (left), SKEWED (right)

(c) Results for $\Sigma = 20$, LINEAR (left), SKEWED (right)

Fig. 3.1 Differences between the results of CMSA and those obtained by applying CPLEX to ILP$_{\text{compl}}$, in percent, concerning the 600 instances of the second benchmark set. Each box shows these differences for the corresponding 10 instances. Negative values indicate that CPLEX obtained a better result than CMSA

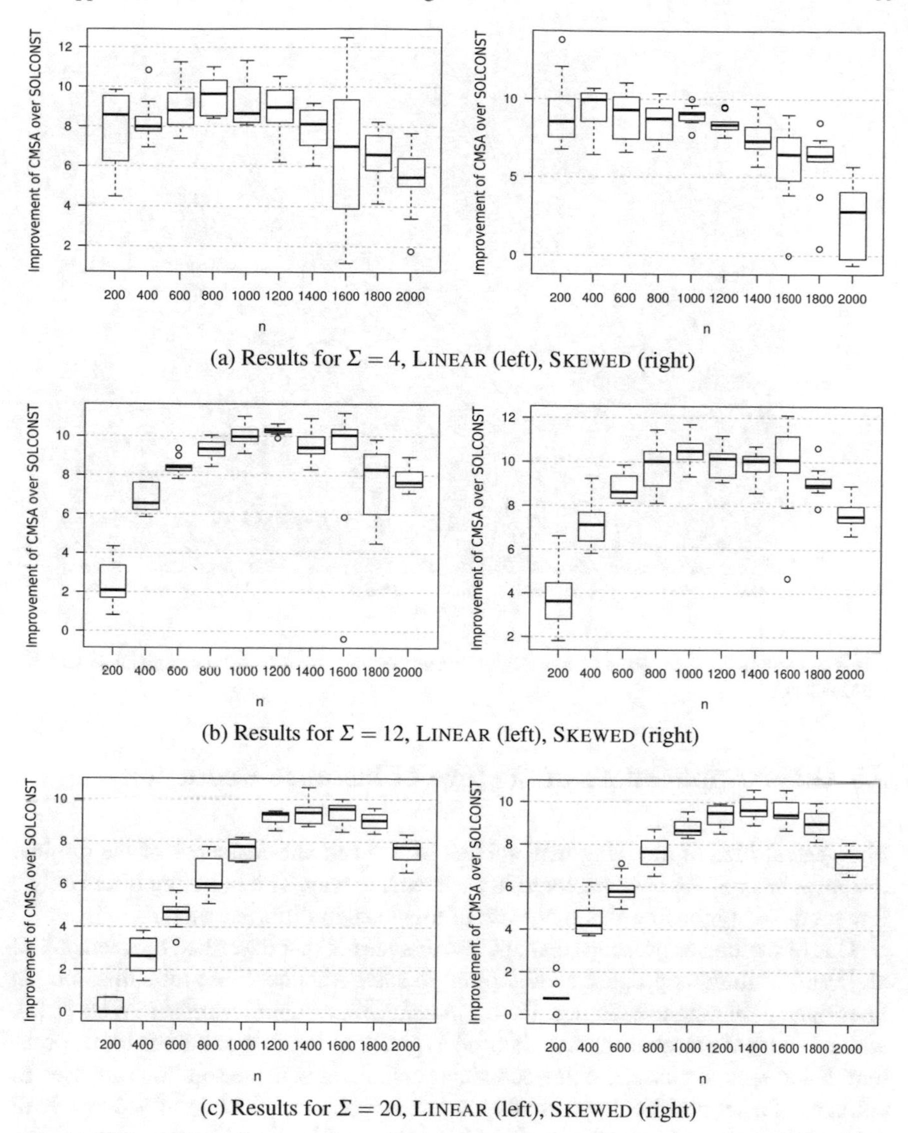

(a) Results for $\Sigma = 4$, LINEAR (left), SKEWED (right)

(b) Results for $\Sigma = 12$, LINEAR (left), SKEWED (right)

(c) Results for $\Sigma = 20$, LINEAR (left), SKEWED (right)

Fig. 3.2 Differences between the results of CMSA and those of SOLCONST, in percent, concerning the 600 instances of the second benchmark set. Each box shows these differences for the corresponding 10 instances

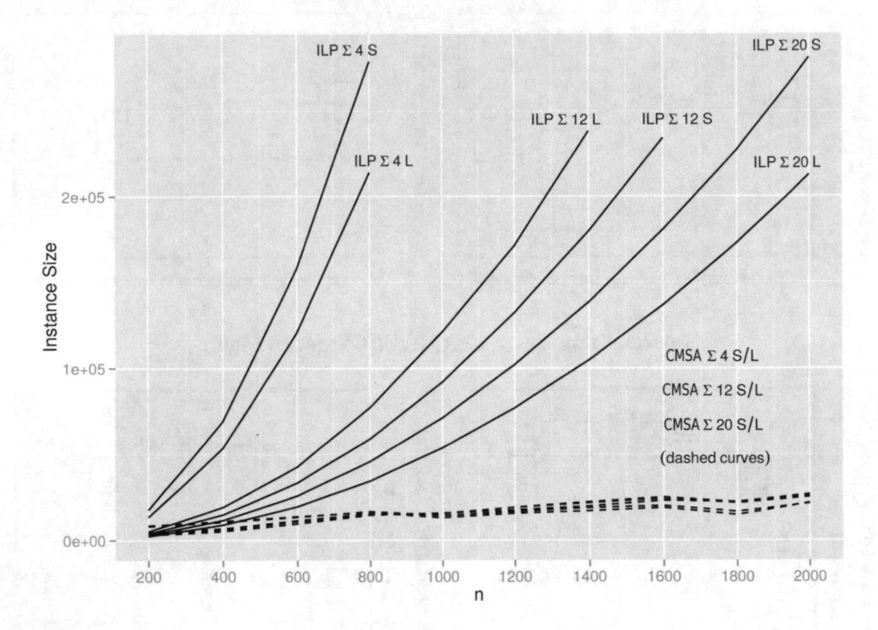

Fig. 3.3 Evolution of the sizes of the original instances and the sub-instances given to CPLEX within CMSA

3.3 Other Applications of the Idea of Instance Reduction

The general idea of applying MIP solvers to reduced sub-instances of the original instances has not yet been explored in a systematic way. The literature offers only a few sporadic approaches that make use of this idea, in different ways.

One of the first approaches that applied this idea is that described by Clements et al. [59]. The authors propose an algorithm to solve a scheduling problem arising in fiber-optic cable manufacturing. First, *Squeaky Wheel Optimization* (SWO) [147], which is a local search technique, is used to generate feasible solutions to the problem. In the second phase, the line-schedules contained in these solutions are used as columns of a set-partitioning formulation of the problem, which is solved by a MIP solver. This process is reported to provide a solution which is at least as good as, but usually better than, the best solution found by SWO.

Applegate et al. [8] and Cook and Seymour [63] tackle the classic traveling salesman problem by means of a two-phase approach. The first phase consists in generating a set of high-quality TSP solutions using a metaheuristic. These solutions are then merged, resulting in a reduced problem instance, which is then solved to optimality by means of an exact solver.

Klau et al. [158] present the following approach for the prize-collecting Steiner tree problem. First, the given problem instance is reduced in such a way that it still contains the optimal solution to the original problem instance. Then, a memetic algorithm is applied to this reduced problem instance. Finally, a MIP solver is applied to find the best solution to the problem instance obtained by merging all solutions of the first and the last population of the memetic algorithm.

3.3.1 The Generate-and-Solve Framework

In [205, 206, 207, 204], the authors described a general algorithm framework labeled Generate-And-Solve (GS). In fact, the CMSA algorithm presented in this chapter can be seen as an instantiation of this framework. The GS framework decomposes the original optimization problem into two conceptually different levels. One of the two levels makes use of a component called *Solver of Reduced Instances* (SRI), in which an exact method is applied to sub-instances of the original problem instance that maintain the conceptual structure of the original instance, that is, any solution to the sub-instance is also a solution to the original instance. At the other level, a metaheuristic component deals with the problem of generating sub-instances that contain high quality solutions. In GS, the metaheuristic component is called the *Generator of Reduced Instances* (GRI). Feedback is provided from the SRI component to the GRI component, for example, by means of the objective function value of the best solution found in a sub-instance. This feedback serves for guiding the search process of the GRI component.

Even though most existing applications of the GS framework are in the context of cutting, packing and loading problems—see, for example, [206, 207, 204, 224, 265]—other successful applications include those to configuration problems arising in wireless networks [71, 70, 225]. Moreover, it is interesting to note that the applications of GS published to date generate sub-instances in the GRI component using either evolutionary algorithms [224, 225] or simulated annealing [70, 265]. Finally, note that in [224] the authors introduced a so-called density control operator in order to control the size of the generated sub-instances. This mechanism can be seen as an additional way of providing feedback from the SRI component to the GRI component.

3.3.2 Solution Merging

The concept of instance reduction can also be found within operators of metaheuristics. In the context of evolutionary algorithms, for example, *solution merging* refers to the idea of exploring the union of two or more solutions by means of a specialized technique. Aggarwal et al. [1] originally suggested such an approach, labeled *optimized crossover*, for the independent set problem. They exploit the fact that a sub-instance obtained by merging two independent sets can efficiently be solved to optimality. Ahuja et al. [2] also make use of this principle within a genetic algorithm for the quadratic assignment problem. They describe a matching-based optimized crossover heuristic that produces an optimized offspring quickly in practice.

The same technique can be applied to other assignment-type problems, as it relies on the structure of the problem rather than the objective function. Borisovsky et al. [39] make use of CPLEX for deriving the best solution from merging two solutions within an evolutionary algorithm. This work is in the context of a supply management problem with lower-bounded demands. Marino et al. [189] present an approach for the graph coloring problem in which a linear assignment problem is solved to optimality in the context of the crossover operator. This is done with the purpose of producing the best possible offspring given two parent solutions.

For general mixed integer programming, Rothberg [260] describes a tight integration of an EA in a branch-and-cut-based MIP solver. In regular intervals, a certain number of iterations of the EA is performed as a branch-and-bound tree node heuristic. Recombination follows the idea of solution merging by first fixing all variables that are common in selected parental solutions. The values of the remaining variables are then determined by applying the MIP solver to the reduced subproblem. Mutation is performed by selecting one parent, fixing a randomly chosen set of variables, and again solving the resulting reduced subproblem by the MIP solver. Since the number of variables to be fixed is a critical parameter, an adaptive scheme is used to control it. This method was integrated into CPLEX in version 10.

The related literature also offers several approaches in which dynamic programming is used for finding the best parent solution in the context of a crossover operator. Yagiura and Ibaraki [300] studied this approach in the context of three permutation problems: the single machine scheduling, the optimal linear arrangement, and the traveling salesman problem. Blum and Blesa [28] present a corresponding approach for the k-cardinality tree problem.

Finally, some authors considered solution merging also from a more theoretical perspective. Cotta and Troya [68] discuss the concept of merging in the context of a framework for hybridizing branch-and-bound with evolutionary algorithms. Based on theoretical considerations, they develop the idea of a dynastically optimal crossover operator. The resulting operator explores the potential of the recombined solutions using branch-and-bound, producing an offspring that is characterized by the best possible combination of its ancestors' features.

Eremeev [89] studies the computational complexity of producing a best possible offspring from two parents for diverse binary problems from a theoretical point of view. The obtained results show that the optimal recombination problem is polynomially solvable for the maximum weight set-packing problem, the minimum weight partitioning problem, and linear Boolean programming problems with at most two variables per inequality. On the other hand, determining an optimal offspring is NP-hard for 0/1 integer programming with three or more variables per inequality, the knapsack problem, set covering, the p-median problem, and others. A recent survey on these results is provided in [90].

Chapter 4
Hybridization Based on Large Neighborhood Search

The type of algorithm addressed in this chapter is based on the following general idea. Given a valid solution to the tackled problem instance—henceforth called the incumbent solution—first, destroy selected parts of it, resulting in a partial solution. Then apply some other, possibly exact, technique to find the best valid solution on the basis of the given partial solution, that is, the best valid solution that contains the given partial solution. Thus, the destroy-step defines a *large neighborhood*, from which a best (or nearly best) solution is determined not by naive enumeration but by the application of a more effective alternative technique. In our examples here, we mainly consider a MIP solver for this purpose, which is applied to a MIP model for the original problem in which the variables corresponding to the given partial solution get respective fixed values preassigned. The motivating aspect for this concept is the same as that described in the context of hybrid algorithms based on instance reduction in Chapter 3: Even though it might be unfeasible to apply a MIP solver to the original problem instance, the solver might be particularly effective in solving the reduced ILP model in which a part of the variables has fixed values.

The outline of this chapter is as follows. After a more detailed description of the general idea, a specific version of a *Large Neighborhood Search* (LNS) algorithm labeled *MIP-based LNS* will be described making use of this concept in an iterated manner. The *Minimum Weight Dominating Set* (MWDS) problem is chosen as a test case in order to demonstrate the application of the algorithm. Obtained results show that the proposed algorithm is superior to the application of a MIP solver to the original problem instance. In another example, MIP-based LNS is integrated in a variable neighborhood search for the *Generalized Minimum Spanning Tree* (GMST) problem, which we already considered in Chapter 2 in the context of incomplete solution representations and decoder-based approaches.

4.1 General Idea

The general optimization approach presented in this chapter is based on the following observation. Imagine it is not possible to apply a MIP solver directly to a considered problem instance due to its size. For smaller instances of the considered problem, however, the MIP approach might be effective. In this case it appears promising to generate a feasible solution of reasonable quality and to try to improve this solution in the following way. First, destroy the solution partially by unfixing some of the variable-value assignments. This might be done in a purely random way, or guided by some heuristic criterion. Afterwards the MIP solver is applied to the problem of finding the best possible solution to the original problem instance that contains the partial solution obtained in the destruction step. Due to the fixed solution parts, this problem's complexity is substantially smaller and the MIP approach should be effective. The whole procedure is iteratively applied to an incumbent solution, which—in the simplest case—corresponds to the best solution found since the start of the algorithm. In such a way, we can still benefit from the advantages of a MIP solver, even in the context of problem instances of large size.

For the remainder of this chapter remember that, given a problem instance $\mathscr{I}$ to a generic problem $\mathscr{P}$, set C represents the *complete set of solution components*, that is, the set of all possible components of which solutions to the problem instance are composed. In the context of this chapter, a valid solution S to $\mathscr{I}$ is represented as a subset of the solution components C, that is, $S \subseteq C$. Moreover, a partial solution S_{partial} is an incomplete solution which can be extended to at least one valid solution, and is, therefore, also a subset of C.

4.1.1 MIP-Based Large Neighborhood Search

A general MIP-based large neighborhood search algorithm, henceforth denoted by MIP-LNS, is shown in Algorithm 19. It works as follows. First, in line 2 of Algorithm 19, an initial incumbent solution S_{cur} is generated in some way, for example, by using a greedy heuristic. Then, at each iteration, the current incumbent solution S_{cur} is partially destroyed, that is, some solution components are removed from S_{cur}. Hereby, the number (percentage) of components to be removed, as well as the way in which these components are selected, are important parameters of the algorithm. The resulting partial solution S_{partial} is fed to a MIP solver; see function ApplyExactSolver(S_{partial}, $t_{\max}$) in line 6 of Algorithm 19. The input parameters for the MIP solver include the current partial solution S_{partial} and a time limit $t_{\max}$. By specifying respective constraints, the MIP solver is forced to include S_{partial} in any considered solution. As output, the MIP solver provides the best complete solution found within the available computation time. This solution, denoted by S'_{opt}, may be optimal among all solutions including S_{partial}. This is in case the time limit $t_{\max}$ was sufficient for the MIP solver to solve the corresponding ILP to optimality. Otherwise, S'_{opt} is simply the best feasible solution found by the MIP solver within

Algorithm 19 MIP-based Large Neighborhood Search (MIP-LNS)

1: **input:** problem instance $\mathscr{I}$, time limit t_{max} for the MIP solver
2: $S_{cur} \leftarrow$ GenerateInitialSolution()
3: $S_{bsf} \leftarrow S_{cur}$
4: **while** CPU time limit not reached **do**
5: $\quad S_{partial} \leftarrow$ DestroyPartially(S_{cur})
6: $\quad S'_{opt} \leftarrow$ ApplyExactSolver($S_{partial}, t_{max}$)
7: $\quad$ **if** S'_{opt} is better than S_{bsf} **then** $S_{bsf} \leftarrow S'_{opt}$
8: $\quad S_{cur} \leftarrow$ ApplyAcceptanceCriterion(S'_{opt}, S_{cur})
9: **end while**
10: **return** s_{bsf}

the given computation time. Finally, the last step of each iteration consists in deciding between S_{cur} and S'_{opt} for the incumbent solution of the next iteration. Options include (1) selecting the better solution among the two, (2) always choosing S'_{opt}, or (3) applying a probabilistic criteria similar to simulated annealing.

4.2 Application to Minimum Weight Dominating Set Problem

In general, *Minimum Dominating Set* (MDS) problems belong to the family of covering problems containing a broad range of NP-hard optimization problems. The aim is to determine a dominating set with minimum cardinality in a given undirected graph $G = (V,E)$. Hereby, a subset of vertices $S \subseteq V$ is called a dominating set if each vertex in V is contained in S or adjacent to some vertex in S. The MDS problem has various practical applications such as, for example, in wireless ad-hoc networks [55, 124]. The most popular model for wireless ad-hoc networks are unit disk graphs (UDGs), in which vertices are randomly scattered in the plane and two vertices are adjacent if and only if the Euclidean distance between them is at most one. However, even in the context of UDGs, the MDS problem is known to be NP-hard [58]. Furthermore, it was shown that the MDS problem for general graphs is equivalent to the set cover problem which is in turn a well-known NP-hard problem [149].

The *Minimum Weight Dominating Set* (MWDS) problem, which is used as a test case for MIP-LNS in this chapter, is one of the possible variants of a MDS problem. In the MWDS, the given graph $G = (V,E)$ comes with a positive integer weight assigned to each vertex. In the context of wireless networks, these weights usually represent mobility characteristics or characteristics of the nodes such as the residual energy or the available bandwidth. The problem of determining a minimum weight dominating set has been shown to be an NP-hard problem; see [107]. Apart from applications in wireless ad-hoc networks [278, 6], the MWDS problem has also been used for modeling, for example, query-focused multidocument summarization [270]. Hereby, a sentence graph G, in which vertices represent sentences, is first generated from a set of documents. If the *cosine similarity* between a pair of sentences is above a given threshold, their corresponding vertices are adjacent in G. With respect to a query p, each sentence s is assigned a weight $w(s)$ based on the distance $1 - cosine(p,s)$.

Due to the inherent computational complexity of the MWDS problem, several approximation algorithms have recently been proposed in the literature. Up to now, the best-known approximation algorithm has a performance ratio of four for UDGs [286, 306]. In this context, note that a ρ-approximation algorithm, with a performance ratio of ρ, runs in polynomial time and always generates a solution whose total weight is at most a factor of ρ worse than the weight of an optimal solution, even in the worst-case. Although the MWDS problem has received considerable attention, there are some limitations with current approaches since most of them are either restricted to UDGs or they impose constraints on the weights. Therefore, researchers have recently turned to metaheuristics as a promising option to obtain high-quality solutions to this problem. Nevertheless, only very few studies have addressed this issue. These include ant colony optimization approaches [148, 234] and a genetic algorithm [234]. However, in particular the genetic algorithm seems not to scale very well and to be quite time-consuming when the problem size increases.

Before starting with the formal description of the MWDS problem, some notations as well as some definitions are introduced in the following. A problem instance of the MWDS problem consists of an undirected graph $G = (V,E)$, where $V = \{v_1,\ldots,v_n\}$ is the set of nodes and $E \subseteq V \times V$ is the set of edges. Each node $v \in V$ has a positive integer weight $w(v)$. For a node $v \in V$, we denote the set of neighbors by $N(v) = \{u \in V \mid (v,u) \in E\}$. $N(v)$ is also known as the *open neighborhood* of v. The *closed neighborhood* $N[v]$ of a node $v \in V$ is defined as $N[v] \leftarrow N(v) \cup \{v\}$. The *degree* $\deg(v)$ of v is the number of v's neighbors, that is, $\deg(v) = |N(v)|$. A *dominating set* $S \subseteq V$ is a subset of the nodes of G such that each node $v \in V \setminus S$ has a neighbor in S. Each node in S is called a *dominator*. A node from $V \setminus S$ is called a *dominatee*. A dominator dominates (covers) itself and all its neighbors.

The MWDS problem can then be stated as follows. A subset $S \in V$ is a valid solution to the MWDS, if and only if S is a dominating set of G. The objective function value $f(S)$ of a valid solution S is defined as the sum of the weights of the vertices in S. The optimization goal consists in finding a dominating set $S^* \subseteq V$ that minimizes the objective function.

The MWDS problem can be formulated in terms of the following ILP model. A binary variable $x_i \in \{0,1\}$ is assigned to each vertex $v_i \in V$ such that $x_i = 1$ if and only if vertex v_i belongs to an optimal solution.

$$\text{minimize} \quad \sum_{i=1}^{n} w(v_i) \cdot x_i \tag{4.1}$$

$$\text{subject to} \quad \sum_{v_j \in N[v_i]} x_j \geq 1 \qquad \forall v_i \in V \tag{4.2}$$

$$x_i \in \{0,1\} \qquad \forall i = 1,\ldots,n \tag{4.3}$$

Algorithm 20 Greedy Heuristic for the MWDS Problem

1: **given:** graph $G = (V, E)$ with node weights $w(v)$, $\forall v \in V$
2: $S \leftarrow \emptyset$
3: $V_{\text{cov}} \leftarrow \emptyset$
4: **while** $V_{\text{cov}} \neq V$ **do**
5: $\quad v^* \leftarrow \text{argmax}_{v \in V \setminus V_{\text{cov}}} \left\{ \frac{\deg(v|V_{\text{cov}})}{w(v)} \right\}$
6: $\quad S \leftarrow S \cup \{v^*\}$
7: $\quad V_{\text{cov}} \leftarrow V_{\text{cov}} \cup N[v^*]$
8: **end while**
9: **return** S

Hereby, inequalities (4.2) ensure that every node $v_i \in V$ is covered, i.e., the subset of nodes indicated by variables $x_i = 1$ forms indeed a dominating set.

The remainder of this section describes the application of MIP-LNS as presented in Algorithm 19 to the MWDS problem. In particular, in the following subsections we will describe the generation of the initial solution, the way in which the incumbent solution is partially destroyed, and the additional constraints that are added to the ILP model outlined above in order to force the inclusion of the nodes present in a partial solution S_{partial}.

4.2.1 Generation of the Initial Solution

The initial incumbent solution S_{cur} is generated using a well-known greedy heuristic for the MWDS problem. For the description of this greedy heuristic, the following additional definition is needed. Henceforth, given a subset S of the nodes V of the input graph G, V_{cov} denotes the set of nodes that are covered by the nodes in S. Moreover,

$$\deg(v \mid V_{\text{cov}}) \leftarrow |N(v) \cap V \setminus V_{\text{cov}}| \qquad \forall v \in V \setminus V_{\text{cov}} \tag{4.4}$$

is the degree of a node $v \in V \setminus V_{\text{cov}}$ when only considering so-far uncovered nodes. The greedy heuristic chooses at each step among the so-far uncovered nodes the node v^* for which $\frac{\deg(v|V_{\text{cov}})}{w(v)}$ is maximal. This greedy heuristic is shown in Algorithm 20.

4.2.2 Partial Destruction of Solutions

The partial destruction of the incumbent solution is one of the crucial design decisions when developing a MIP-LNS method. In the context of the MWDS problem it was decided that, at each iteration, a certain dynamically determined percentage $perc_{\text{dest}}$ of the nodes present in the current incumbent solution S_{cur} are to be removed. Moreover, the number of nodes to be removed is lower-bounded by three.

More specifically, the exact number d of nodes to be deleted from S_{cur} at each iteration is:

$$d \leftarrow \max\left\{3, \left\lfloor \frac{perc_{\text{dest}} \cdot |S_{\text{cur}}|}{100} \right\rfloor\right\}. \tag{4.5}$$

The next question concerns the way in which the d nodes to be removed are chosen from S_{cur}. For this purpose, the following two options are considered. The first one, denoted by $type_{\text{dest}} = 0$, consists in selecting d nodes from S_{cur} uniformly at random. In the second option, denoted by $type_{\text{dest}} = 1$, the d nodes are selected according to probabilities depending on their weight value and their degree concerning input graph G. The probability of a node to be chosen for deletion is:

$$\mathbf{p}(v) \leftarrow \frac{w(v)/\deg(v)}{\sum_{v' \in S_{\text{cur}}} w(v')/\deg(v')} \qquad \forall v \in S_{\text{cur}}. \tag{4.6}$$

Finally, in order to obtain a dynamically changing value for $perc_{\text{dest}}$, the following mechanism for the adaptation of $perc_{\text{dest}}$ was employed at the end of each iteration of MIP-LNS. First, the initial value of $perc_{\text{dest}}$ at the start of algorithm MIP-LNS is set to a fixed lower bound $perc^{\text{l}}_{\text{dest}}$. Then, at each iteration, if S'_{opt} is better than S_{cur}, the value of $perc_{\text{dest}}$ is set back to the lower bound $perc^{\text{l}}_{\text{dest}}$, while otherwise, the value of $perc_{\text{dest}}$ is incremented by five, that is,

$$perc_{\text{dest}} \leftarrow perc_{\text{dest}} + 5. \tag{4.7}$$

However, if the value of $perc_{\text{dest}}$ exceeds a fixed upper bound $perc^{\text{u}}_{\text{dest}}$, the value of $perc_{\text{dest}}$ is set back to the lower bound $perc^{\text{l}}_{\text{dest}}$. Obviously it must hold that $0 \leq perc^{\text{l}}_{\text{dest}} \leq perc^{\text{u}}_{\text{dest}} \leq 100$. The lower and the upper bound are parameters of the algorithm.

In summary, the employed mechanism for the partial destruction of solutions has three parameters for which well-working values must be found: the destruction type $type_{\text{dest}} \in \{0,1\}$, and the lower and upper bounds—that is, $perc^{\text{l}}_{\text{dest}}$ and $perc^{\text{u}}_{\text{dest}}$—of the percentage of nodes to be removed from S_{cur}. Remember that the process of partially destroying a solution results in a partial solution S_{partial}.

4.2.3 Application of the MIP Solver

Given a partial solution S_{partial}, and a computation time limit $t_{\max}$ for the MIP solver, the solver is applied to the ILP model for the MWDS problem as outlined in (4.1)–(4.3), extended with the following additional constraints:

$$x_i = 1 \qquad \forall v_i \in S_{\text{partial}}. \tag{4.8}$$

These constraints ensure that the nodes present in S_{partial} must appear in the solution generated by the MIP solver.

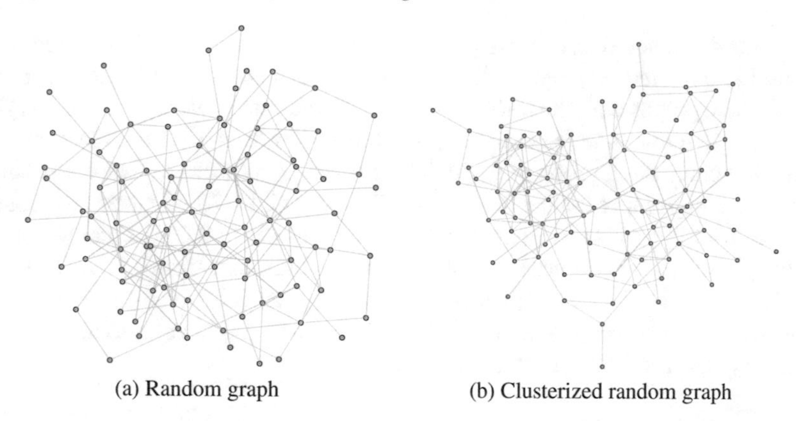

(a) Random graph (b) Clusterized random graph

Fig. 4.1 Examples of the types of generated graphs. Displayed graphs consist of 100 nodes

4.2.4 Experimental Evaluation

The proposed application of MIP-LNS to the MWDS problem was implemented in ANSI C++ using GCC 4.7.3 for compiling the software. Moreover, both the complete ILP model and the extended ILP models within MIP-LNS were solved with IBM ILOG CPLEX 12.1. The experimental evaluation was conducted on a cluster of 32 PCs with Intel Xeon CPU X5660 CPUs of 2 nuclei of 2.8 GHz and 48 GB of RAM. The following algorithms were considered for the comparison:

1. GREEDY: the greedy approach from Algorithm 20.
2. $\text{ILP}_{\text{compl}}$: the application of CPLEX to the complete ILP for each considered problem instance.
3. MIP-LNS: the LNS algorithm outlined before.

With the aim of testing the proposed algorithm in different scenarios we created two different sets of graphs: random graphs (RG) and clusterized random graphs (CRG). In each case, graphs of different properties—for what concerns, for example, the density—and different sizes were created. In particular, for each type we generated graphs with $|V| \in \{100, 1000, 5000, 10000\}$ nodes. Ten random graphs were created for each combination between graph characteristics and size.

The graphs were generated adding edges between nodes at random with a predefined probability. This probability controls the density of the graph. As mentioned above, two different types of random graphs were generated: (1) standard random graphs where any two nodes have the same probability of being connected and (2) clusterized random graphs, where nodes are divided into two groups and the probabilities of being connected in each cluster and between clusters are different. Regarding the standard random graphs, we selected the probability of connecting two nodes (p) from $\{0.03, 0.04, 0.05\}$ in the context of graphs of size 100 and from $\{0.01, 0.03, 0.05\}$ for the remaining graph sizes. Henceforth we refer to the resulting graph densities as *low, medium, and high*. The case of two-cluster graphs is more complex. We used two different probabilities—henceforth labeled low and high—for the generation of these graphs. In the case of $n = 100$, low and high correspond to 0.06 and 0.1, and for the remaining graph sizes to 0.02 and 0.1, respectively. Then, graphs with the following three configurations were generated: (1) both clusters have a low inner connection probability (low,low), (2) both clusters have a high inner connection probability (high,high), and (3) one cluster has a low and the other one a high inner connection probability (low,high). In all cases the probability of connecting nodes from different clusters was set to 0.005. Finally, note that the weight of each node was chosen uniformly at random from $\{1,\ldots,100\}$. Figures 4.1a and 4.1b show examples of random graphs and clusterized random graphs, respectively. Node weights are not indicated in these graphics. In total, the benchmark set consists of 240 benchmark instances.

4.2.4.1 Tuning of MIP-LNS

MIP-LNS has several parameters for which well-working values must be found: (1) the destruction type $type_{\text{dest}} \in \{0,1\}$, (2) the lower and upper bounds—that is, $perc^{\text{l}}_{\text{dest}}$ and $perc^{\text{u}}_{\text{dest}}$—of the percentage of nodes to be removed from the current incumbent solution S_{cur}, and (3) the maximum time $t_{\max}$ (in seconds) allowed for CPLEX per application to an instance extended with additional constraints. The last parameter is necessary, because even when applied to problem instances with a reduced search space (obtained by adding additional constraints), CPLEX might still need too much computation time for solving such instances to optimality. In any case, CPLEX always returns the best feasible solution found within the given computation time.

The automatic configuration tool irace [179] was used for the fine-tuning of the parameters. In fact, irace was applied to tune MIP-LNS separately for each instance size from $|V| \in \{100,500,1000,10000\}$. In total, 12 *training instances* for each of the four different instance sizes were generated for tuning: six instances concerning random graphs and further six instances concerning clusterized random graphs. The six instances per graph type are composed of two instances per considered graph density in the case of random graphs, and two instances per density combination in the case of clusterized random graphs. The tuning process for each instance size was given a budget of 1000 runs of MIP-LNS, where each run was terminated after $|V|/10$ CPU seconds. The following parameter value ranges were chosen concerning the parameters of MIP-LNS:

- $type_{\text{dest}} \in \{0,1\}$, where a setting of zero indicates a random procedure and one a heuristically biased random procedure.
- For the bounds of the node removal probabilities, we considered the settings $(perc^{\text{l}}_{\text{dest}}, perc^{\text{u}}_{\text{dest}}) \in \{(10,10), (20,20), (30,30), (40,40), (50,50), (60,60), (70,70), (80,80), (90,90), (10,30), (10,50), (30,50), (30,70)\}$. In those cases in which both bounds have the same value, the percentage of deleted nodes is always the same. Only when the lower bound is lower than the upper bound, a dynamically changing percentage is used.

Table 4.1 Parameter settings produced for MIP-LNS by irace for the four different instance sizes

$\lvert V\rvert$	$type_{dest}$	$(perc^{l}_{dest}, perc^{u}_{dest})$	t_{max}
100	1	(60,60)	2.0
1000	0	(90,90)	10.0
5000	1	(50,50)	5.0
10000	1	(40,40)	10.0

- The setting of t_{max} (in seconds) was made dependent on the instance size:
  1. If $|V| = 100$: $t_{max} \in \{0.5, 1.0, 1.5, 2.0\}$
  2. If $|V| = 1000$: $t_{max} \in \{1.0, 4.0, 7.0, 10.0\}$
  3. If $|V| = 5000$: $t_{max} \in \{5.0, 10.0, 15.0, 20.0\}$
  4. If $|V| = 10000$: $t_{max} \in \{10.0, 20.0, 30.0, 40.0\}$

The four applications of irace produced the four configurations for MIP-LNS shown in Table 4.1. The following tendencies can be observed. First, the heuristically biased random choice of nodes to be deleted for the partial destruction of solutions seems, in general, to be beneficial. Second, a dynamically changing probability for the node removal (depending on the search history) does not seem to add to the success of the algorithm, at least not in the case of the MWDS problem.

4.2.4.2 Results

Numerical results are shown in Table 4.3a concerning standard random graphs, and in Table 4.3b concerning clusterized random graphs. The results are presented in these tables in terms of averages (means) over 10 random instances of the same characteristics. Each algorithm included in the comparison was applied once to each problem instance.

The structure of the two tables is as follows. The first column provides the instance size in terms of the number of nodes, whereas the second column indicates the density of the corresponding graphs. The third and fourth columns contain the results of GREEDY, in terms of the obtained average solution quality (column **mean**) and the average computation time (in seconds) needed to produce these results (column **time**). The next three table columns are dedicated to the presentation of the results provided by solving the ILP model $\text{ILP}_{\text{compl}}$. The first one of these columns shows the values of the best solutions found within $|V|/10$ CPU seconds. The second column lists computation times in seconds, more specifically, two values in the form X/Y, where X corresponds to the time at which CPLEX was able to find the first valid solution and Y to the time at which CPLEX found the best—possibly optimal—solution within the time limit. Finally, the third column of those dedicated to $\text{ILP}_{\text{compl}}$ shows the average optimality gap, which refers to the relative difference between the value of the best valid solution and the current lower bound at the time of stopping a run. Finally, the last four columns of each table are dedicated to the presentation of the results obtained by MIP-LNS. The first one of these columns

Table 4.2 Numerical results for the MWDS problem

$\|V\|$	**density**	Greedy		ILP$_{\text{compl}}$			Mip-Lns			
		mean	**time**	**mean**	**time**	**gap**	**mean**	**time**	**ILP time**	**# iters**
	low	1202.0	<0.1	968.0	0/0	0.0%	968.0	<0.1	99%	6228.9
100	medium	1039.4	<0.1	804.3	0/0	0.0%	804.3	<0.1	100%	4853.1
	high	813.5	<0.1	644.6	0/0	0.0%	644.6	<0.1	100%	4735.5
	low	3472.3	<0.1	2906.0	0/12	0.0%	2906.0	13.7	100%	383.0
1000	medium	700.8	<0.1	605.1	0/61	0.8%	606.0	33.1	100%	31.2
	high	362.0	0.1	318.9	0/56	1.2%	318.7	26.2	100%	26.2
	low	1758.6	2.1	1570.9	4/499	13.6%	1530.9	264.2	99%	97.9
5000	medium	330.8	5.7	304.8	4/499	22.7%	294.8	225.9	99%	85.3
	high	163.5	6.5	152.7	5/500	25.9%	147.2	224.8	99%	76.2
	low	1213.3	20.6	1147.2	28/779	22.4%	1085.6	561.4	98%	103.6
10000	medium	249.4	28.7	238.2	24/896	32.7%	225.1	468.8	97%	72.9
	high	120.9	30.6	118.5	31/978	39.2%	110.8	511.4	97%	45.4

(a) Results for random graphs

$\|V\|$	**density**	Greedy		ILP$_{\text{compl}}$			Mip-Lns			
		mean	**time**	**mean**	**time**	**gap**	**mean**	**time**	**ILP time**	**# iters**
	(low,low)	1195.1	<0.1	954.7	0/0	0.0	954.7	<0.1	99%	5439.0
100	(low,high)	922.0	<0.1	759.2	0/0	0.0	759.2	<0.1	100%	5407.0
	(high,high)	795.9	<0.1	637.7	0/0	0.0	637.7	<0.1	100%	4801.6
	(low,low)	2516.4	<0.1	2119.0	0/35	0.2	2119.0	23.3	100%	89.3
1000	(low,high)	1588.5	0.1	1340.4	0/6	0.0	1340.4	30.2	100%	118.4
	(high,high)	303.4	0.1	269.3	0/35	0.3	269.3	21.9	100%	45.1
	(low,low)	1227.9	3.9	1100.3	4/497	14.8	1074.9	272.3	99%	95.3
5000	(low,high)	815.3	8.6	716.7	5/500	11.0	708.5	257.2	98%	128.3
	(high,high)	152.8	6.5	140.9	6/500	24.9	136.6	124.8	99%	78.6
	(low,low)	883.8	21.6	836.7	25/863	24.0	790.2	636.2	98%	98.0
10000	(low,high)	583.0	45.9	547.2	26/948	23.0	516.7	770.8	96%	70.2
	(high,high)	115.7	31.1	110.8	38/989	38.9	105.3	359.3	97%	42.5

(b) Results for clusterized random graphs

provides the average solution qualities obtained within $|V|/10$ CPU seconds. The second column indicates the average computation time needed by Mip-Lns to find the best solution of a run. The third column shows the percentage of the total computation time of a run that was spent by applications of the MIP solver. Finally, the fourth column shows the average number of algorithm iterations executed within the allowed computation time. The best result of each row is marked by a gray background.

The analysis of the results permits us to draw the following conclusions:

- Hardly any differences can be observed in the relative performance of the algorithms concerning the two different instance types, that is, standard random graphs and clusterized random graphs. Therefore, all the following statements hold for both types of graphs.

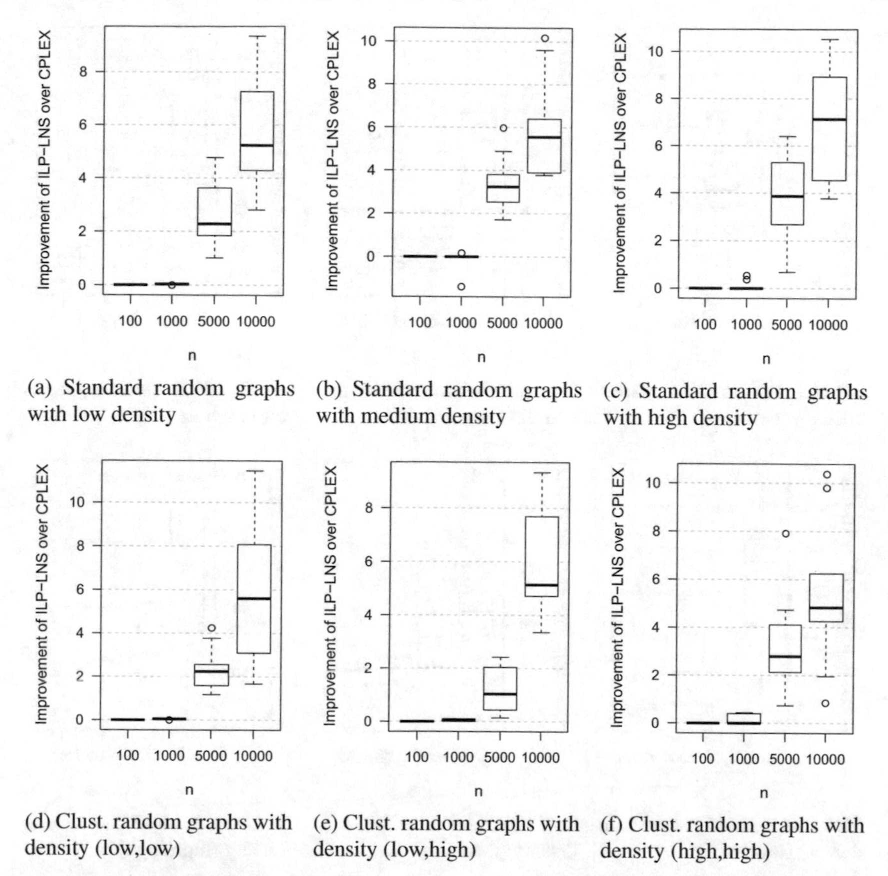

(a) Standard random graphs with low density

(b) Standard random graphs with medium density

(c) Standard random graphs with high density

(d) Clust. random graphs with density (low,low)

(e) Clust. random graphs with density (low,high)

(f) Clust. random graphs with density (high,high)

Fig. 4.2 Differences between the results of MIP-LNS and those obtained by applying CPLEX to $\text{ILP}_{\text{compl}}$ (in percent) concerning all the instances from the considered benchmark set. Each box shows these differences for the corresponding 10 instances. Negative values indicate that CPLEX obtained a better result than MIP-LNS

- Concerning the application of CPLEX to $\text{ILP}_{\text{compl}}$, the graph density appears to have a strong influence on the problem difficulty. With increasing graph densities—considering graphs of the same number of nodes—the average optimality gaps obtained by CPLEX also increase.
- For rather small instances, CPLEX and MIP-LNS provide—with the exception of standard random graphs, $|V| = 1000$, medium density—exactly the same results. Moreover, both methods require a comparable amount of computation time for doing so.
- Concerning larger instances, MIP-LNS clearly outperforms CPLEX concerning graphs of all densities.

In summary, we can state that MIP-LNS is competitive with the application of CPLEX to the original ILP model when the size of the input instances is rather

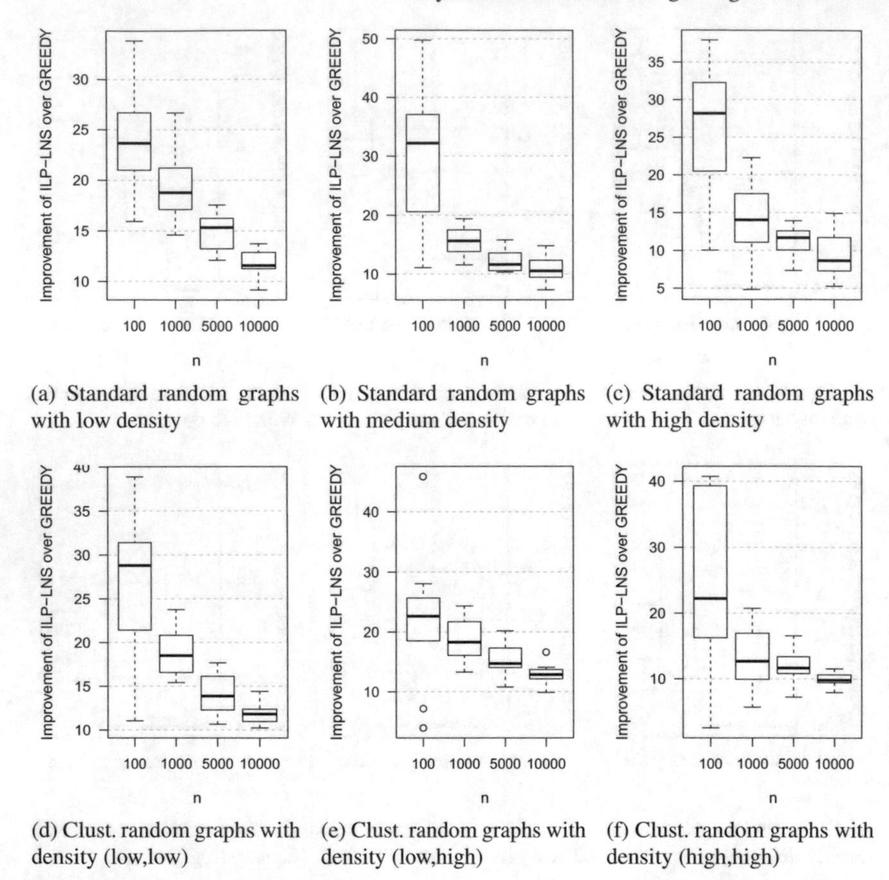

(a) Standard random graphs with low density

(b) Standard random graphs with medium density

(c) Standard random graphs with high density

(d) Clust. random graphs with density (low,low)

(e) Clust. random graphs with density (low,high)

(f) Clust. random graphs with density (high,high)

Fig. 4.3 Differences between the results of MIP-LNS and those obtained by GREEDY (in percent) concerning all the instances from the considered benchmark set. Each box shows these differences for the corresponding 10 instances

small. The larger the size of the tackled problem instances, however, the greater is the advantage of MIP-LNS over the other algorithms. The validity of these statements can be conveniently observed in the graphics of Figure 4.2 and Figure 4.3.

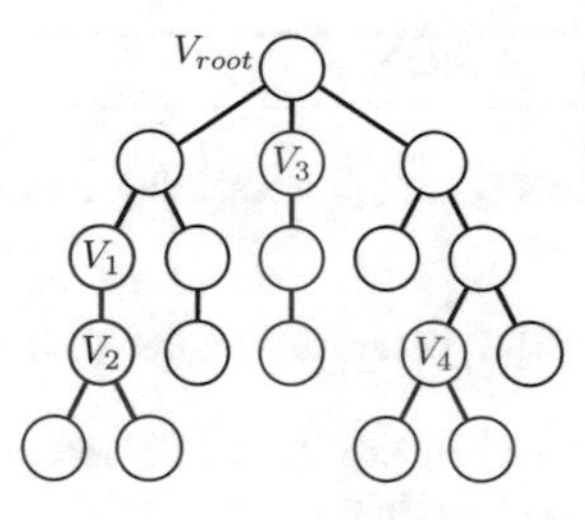

Fig. 4.4 GSON neighborhood for the GMST problem: Selection of subtrees to be optimized via the MIP solver (from [142])

4.3 Application to the Generalized Minimum Spanning Tree Problem

As an additional example, we extend COMB-VNS, the variable neighborhood search for the generalized minimum spanning tree (GMST) problem presented in Section 2.2 to illustrate incomplete solution representations and decoders, by an additional MIP-based LNS. As we will see, this extension further improves the overall performance of the VNS. This presentation is based on [142]. For a better understanding of this section we recommend the reader to first study Chapter 2 of this book, which introduces the GMST problem and the above-mentioned VNS approach.

4.3.1 Global Subtree Optimization Neighborhood

The *Global Subtree Optimization Neighborhood* (GSON) follows the above described principle of partially destroying a solution and re-optimizing the destroyed part by means of a MIP solver.

Let us consider a current GMST solution $S = (P,T)$ with its corresponding global spanning tree $S^g = (V^g, T^g)$ on the global graph G^g as defined in Section 2.2, i.e., for each edge $(u,v) \in T$ with $u \in V_i \wedge v \in V_j$, there exists a global edge $(V_i, V_j) \in T^g$. After rooting S^g at a randomly chosen cluster V_{root}, we perform a depth-first search to determine all subtrees $Q_1, \ldots, Q_k$ containing at least $N_{\min}$ and no more than $N_{\max}$ clusters, with $N_{\min}$ and $N_{\max}$ being strategy parameters. Figure 4.4 shows an example for this subtree selection with $N_{\min} = 3$ and $N_{\max} = 4$ yielding subtrees $Q_1, \ldots, Q_4$ rooted at $V_1, \ldots, V_4$.

A move to a neighbor solution in GSON is performed by globally re-optimizing one subtree Q_i, that is, we consider the complete subgraph induced by the clusters and nodes of Q_i as an independent GMST problem and optimally solve this problem by a MIP solver. The obtained new subtree is then connected to the unchanged remainder of the current solution in the best possible way. The latter can

Algorithm 21 Exploration of GSON

given: solution S
$V_1,\ldots,V_k$ = roots of the subtrees $Q_1,\ldots,Q_k$ containing at least $N_{\min}$ and no more than $N_{\max}$ clusters
for all $i \leftarrow 1,\ldots,k$ **do**
  remove the edge (parent of V_i, V_i) {separate subtree Q_i from S}
  optimize Q_i via MIP solver
  reconnect Q_i to S in a best possible way {as GEEN reconnection mechanism}
  if current solution better than best **then**
    save current solution as best
  end if
  restore initial solution
end for
return best solution

be achieved by inspecting all global edges connecting both components, as is also done in GEEN. Algorithm 21 shows this large neighborhood search in pseudo-code.

The computational complexity of GSON is hard to state due to the used MIP-solver and obviously depends on $N_{\min}$ and $N_{\max}$. The number of subtrees to be considered is bounded below by 0 and above by $\lfloor \frac{r}{N_{\max}} \cdot (N_{\max} - N_{\min} + 1) \rfloor$.

As MIP model, the following *local-global formulation* from Pop [232] is used:

$$\text{minimize} \quad \sum_{e \in E} c_e x_e \tag{4.9}$$

$$\text{subject to} \quad \sum_{v \in V_k} z_v = 1 \qquad \forall k = 1,\ldots,r \tag{4.10}$$

$$\sum_{e \in E} x_e = r - 1 \tag{4.11}$$

$$\sum_{e=(u,v) \mid u \in V_i, v \in V_j} x_e = y_{ij} \qquad \forall i,j = 1,\ldots,r,\ i \neq j \tag{4.12}$$

$$\sum_{e=(u,v) \mid u \in V_i} x_e \leq z_v \qquad \forall i = 1,\ldots,r,\ \forall v \in V \setminus V_i \tag{4.13}$$

$$y_{ij} = \lambda_{kij} + \lambda_{kji} \qquad \forall i,j,k = 1,\ldots,r,\ i \neq j,\ i \neq k \tag{4.14}$$

$$\sum_{j \in \{1,\ldots,r\} \setminus \{i\}} \lambda_{kij} = 1 \qquad \forall i,k = 1,\ldots,r,\ i \neq k \tag{4.15}$$

$$\lambda_{kkj} = 0 \qquad \forall j,k = 1,\ldots,r,\ j \neq k \tag{4.16}$$

$$\lambda_{kij} \geq 0 \qquad \forall i,j,k = 1,\ldots,r,\ i \neq j,\ i \neq k \tag{4.17}$$

$$x_e,\ z_v \geq 0 \qquad \forall e \in E, \forall v \in V \tag{4.18}$$

$$y_{lr} \in \{0,1\} \tag{4.19}$$

Variables z_v indicate the connected nodes and variables x_e the used edges from G. Variables y_{ij} specify the used global edges, and more specifically, variables $\lambda_{k,i,j}$ are set to one if cluster V_j is the predecessor of cluster V_i when rooting the tree at cluster V_k. Equations (4.10) guarantee that only one node is selected per cluster.

Table 4.3 Impact of different settings for GSON's $N_{\min}$ and $N_{\max}$ parameters on final results of COMB*-VNS

Instances	$N_{\min}=3, N_{\max}=4$		$N_{\min}=5, N_{\max}=6$		$N_{\min}=7, N_{\max}=8$		$N_{\min}=3, N_{\max}=8$	
	$c(T)$	$\sigma(c(T))$	$c(T)$	$\sigma(c(T))$	$c(T)$	$\sigma(c(T))$	$c(T)$	$\sigma(c(T))$
Group. Eucl. 125	133.8	0.00	133.8	0.00	133.8	0.00	133.8	0.00
Group. Eucl. 500	585.3	1.01	585.0	1.32	584.4	1.50	584.8	1.32
Group. Eucl. 600	87.9	0.00	87.9	0.00	88.4	0.39	87.9	0.00
Group. Eucl. 1280	318.4	1.58	318.3	1.78	319.5	1.51	319.6	1.97
Rand. Eucl. 250	2208.6	31.66	2201.8	23.30	2198.6	20.56	2212.3	33.31 1
Rand. Eucl. 400	595.3	0.00	595.3	0.00	618.3	24.62	621.6	25.02
Rand. Eucl. 600	530.5	7.53	537.0	10.2	579.7	42.90	562.2	20.79
Non-Eucl. 200	41.0	0.00	41.0	0.00	41.0	0.00	41.0	0.00
Non-Eucl. 500	149.1	6.54	148.6	4.27	148.4	6.28	148.5	3.84
Non-Eucl. 600	16.7	1.42	16.1	1.24	16.3	1.42	17.0	1.66
kroa150	9815.0	0.00	9815.0	0.00	9815.0	0.00	9815.0	0.00
krob200	11244.0	0.00	11244.0	0.00	11244.0	0.00	11244.0	0.00
ts225	62268.2	0.38	62268.5	0.51	62269.4	4.68	62268.3	0.47
gil262	942.1	0.57	942.3	1.02	942.0	0.00	942.1	0.57
pr299	20320.9	12.98	20322.6	14.67	20316.1	0.55	20316.8	2.07
rd400	5947.8	10.25	5943.6	9.69	5943.9	9.99	5945.8	10.57
gr431	1033.0	0.00	1033.0	0.18	1033.0	0.00	1033.0	0.18
pcb442	19693.1	49.05	19702.8	52.11	19712.4	53.58	19699.6	53.78

Equation (4.11) forces the solution to contain exactly $r-1$ edges, while (4.12) allow them only between nodes of clusters which are connected in the global graph. Inequalities (4.13) ensure that edges only connect nodes v for which $z_v = 1$. For each $k = 1,\ldots,r$, constraints (4.14) and (4.16) force variables λ_{kij} to represent a directed spanning tree (that is, an arborescence) rooted in V_k: Equations (4.14) ensure the selection of a global edge (i, j) iff i is parent of j or j is parent of i in an arborescence rooted in V_k. Constraints (4.15) guarantee that each cluster except root k has exactly one parent, while (4.16) ensure that root k has no parents. For a deeper discussion of this model and a comparison to other formulations, we refer to [232].

4.3.2 Results of COMB*-VNS: The GSON-Enhanced VNS

We use COMB-VNS exactly as proposed in Section 2.2 but include GSON as a fourth VND neighborhood and call this new variant COMB*-VNS. The time limit was again 600 seconds per run, and 30 runs were performed for each instance.

Important parameters for GSON are the minimum and maximum number of clusters of the subtrees to be optimized via the MIP solver, i.e., $N_{\min}$ and $N_{\max}$. Table 4.3 shows the impact of different settings on the final objective values for a subset of all benchmark instances. In general, results are ambiguous. For Random Euclidean instances, a tendency towards smaller sizes yielding better results is noticeable. An explanation might be that solving these smaller neighborhoods is faster and therefore more VND/VNS iterations can be performed. On some other instances, however, the search process benefits from larger GSON neighborhoods. We decided to

Table 4.4 Average final objective values and standard deviations of COMB*-VNS in comparison to COMB-VNS for the instance sets from Ghosh and Hu. Three instances are considered per type and size

instance set	COMB-VNS		COMB*-VNS	
	$\overline{c(T)}$	$\sigma(c(T))$	$\overline{c(T)}$	$\sigma(c(T))$
Grouped Eucl. 125	141.1	0.00	141.1	0.00
	133.8	0.00	133.8	0.00
	141.4	0.00	141.4	0.00
Grouped Eucl. 500	568.6	0.59	567.4	0.57
	581.0	1.39	585.0	1.32
	587.9	4.07	583.7	1.82
Grouped Eucl. 600	84.8	0.27	84.6	0.11
	87.9	0.05	87.9	0.00
	88.5	0.00	88.5	0.00
Grouped Eucl. 1280	321.8	2.41	315.9	1.91
	316.3	0.83	318.3	1.78
	334.3	2.13	329.4	1.29
Random Eucl. 250	2336.9	34.23	2300.9	40.27
	2304.1	47.95	2201.8	23.30
	2049.8	15.29	2057.6	31.58
Random Eucl. 400	625.4	14.59	615.3	10.8
	595.3	0.14	595.3	0.00
	588.8	7.40	587.3	0.00
Random Eucl. 600	443.5	0.00	443.5	0.00
	535.2	12.20	537.0	10.2
	479.9	26.55	469.0	11.9
Non-Eucl. 200	71.6	0.02	71.6	0.00
	41.0	0.00	41.0	0.00
	52.8	0.00	52.8	0.00
Non-Eucl. 500	173.4	8.40	152.5	3.69
	154.6	6.55	148.6	4.27
	180.1	3.67	166.1	2.89
Non-Eucl. 600	15.9	2.07	15.6	1.62
	17.6	1.75	16.1	1.24
	15.1	0.22	16.0	1.66

set $N_{\min} = 5$ and $N_{\max} = 6$ in order to obtain a balanced default behavior. More detailed results for different settings can be found in [142].

Table 4.4 lists the average final solution qualities of COMB*-VNS in comparison to COMB-VNS for the instance sets of Ghosh and Hu. In total, COMB*-VNS yielded for 15 instances on average better solutions than COMB-VNS, while COMB-VNS was superior in only five cases. The same objective values were achieved in 10 cases. Overall, a Wilcoxon rank test indicates with an error probability of less than 1% that COMB*-VNS is able to find better solutions than COMB*-VNS, and thus clearly documents the benefits of the added GSON. Furthermore, the standard deviations of COMB*-VNS are smaller than those of COMB-VNS in most cases, which might give an indication of a slightly improved robustness.

Table 4.5 Average final objective values and standard deviations of COMB*-VNS in comparison to COMB-VNS for Feremans' TSPlib-based geographically clustered instances

instances	COMB-VNS		COMB*-VNS	
	$\overline{c(T)}$	$\sigma(c(T))$	$\overline{c(T)}$	$\sigma(c(T))$
gr137	329.0	0.00	329.0	0.00
kroa150	9815.0	0.00	9815.0	0.00
d198	7044.0	0.00	7044.0	0.00
krob200	11244.0	0.00	11244.0	0.00
gr202	242.0	0.00	242.0	0.00
ts225	62280.5	16.28	62268.5	0.51
pr226	55515.0	0.00	55515.0	0.00
gil262	943.2	1.63	943.2	1.02
pr264	21890.5	5.92	21886.5	1.78
pr299	20347.4	28.09	20322.6	14.67
lin318	18511.2	9.70	18506.8	11.58
rd400	5955.0	7.57	5943.6	9.69
fl417	7982.0	0.00	7982.0	0.00
gr431	1033.0	0.25	1033.0	0.18
pr439	51849.7	39.30	51847.9	40.92
pcb442	19729.3	50.90	19702.8	52.11

The results for the TSPlib-based geographically clustered instances are shown in Table 4.5. Here, COMB*-VNS performed never worse than COMB-VNS and was superior in seven out of 16 instances.

Average improvement rates of GSON are about 15% and average relative gains about 5%. Both are significantly lower than those of the other neighborhood searches in COMB*-VNS (compare Tables 2.5 and 2.6) but appear still reasonable, especially when considering that GSON is applied to solutions that are already locally optimal w.r.t. all the other neighborhood structures.

4.4 Other Applications of the Idea of Large Neighborhood Search

The *large neighborhood* used in the context of the MWDS example presented in this chapter—achieved by deleting the values of a portion of the decision variables concerning the incumbent solution—is a rather straightforward example. Obviously, the selection of the variables that are subject to optimization and those whose value remains fixed, respectively, plays a crucial role: The number of free variables directly determines the size of the large neighborhood. The smaller the neighborhood, the more unlikely it is that the MIP solver yields an improved solution. On the contrary, if the neighborhood is too large, the MIP solver might require excessive times or might not yield a meaningful solution within the allowed time. Therefore, a strategy for dynamically adapting the number of free variables—as in the case of the

MWDS application—is sometimes used. Furthermore, the variables to be optimized might be selected either purely at random or in a more sophisticated, guided way by considering the variables' potential impact on the objective function and their relatedness. For example, Mitrović-Minić and Punnen [199] describe such an approach for solving general MIP problems, called *variable intensity local search*, and tested it specifically on the generalized assignment problem and its multiresource variant. Moreover, adaptive versions of LNS have been proposed in which a number of destroy and repair operators[1] compete among each other. Operators that perform better have a higher probability of being used again in future iterations. Examples of such adaptive LNS versions can be found, for example, in [258, 100, 218].

A more problem-specific example of a large neighborhood was described by Prandtstetter and Raidl in the context of the car sequencing problem [235]. In this problem, the goal is to find a cost-effective arrangement—that is, a permutation—of commissioned cars along a production line. Each car requires particular components to be installed at different working bays along the assembly line. The objective concerns the smoothing of the workload at the working bays. More formally, no more than l_c cars are allowed to require component c in any subsequence of m_c consecutive cars, and violations of this constraint are penalized by additional cost terms added to the objective function. The authors describe a generalized VNS approach (for an introduction to VNS see Section 1.2.6) that makes use of eight different types of neighborhood structures. Besides the more standard move and swap neighborhoods, more powerful κ-exchange neighborhoods are considered: A set of κ cars is selected either uniformly at random or by means of a greedy strategy giving preference to cars involved in conflicts, thus inducing higher costs. These cars are then released from their current positions and reassigned in an optimal way by solving a corresponding MIP. The number κ of cars to be reassigned is varied within the variable neighborhood search, starting with a small value, which implies small neighborhoods, and increasing it up to a pre-defined upper bound when no improved solution can be found. To avoid too long running times for larger κ, the MIP solver is aborted when a certain time limit is exceeded and the so far best feasible solution (if any) is considered. Empirical investigations have shown that the utilization of the MIP-neighborhood substantially improves the overall solution quality.

An alternative way to define large neighborhoods is to reduce the search space of the original problem instance by adding additional constraints. Among the more generally applicable approaches are so-called *local branching constraints* [98], which—in their basic version—are suited for MIPs with binary variables $(x_1,\ldots,x_n) \in \{0,1\}^n$. Given an incumbent solution $\bar{x} = (\bar{x}_1,\ldots,\bar{x}_n)$, the local branching neighborhood is obtained by adding the following constraint to the original MIP:

$$\Delta(x,\bar{x}) \leftarrow \sum_{j\in\bar{S}}(1-x_j) \quad + \sum_{j\in\{1,\ldots,n\}\setminus\bar{S}} x_j \leq k, \tag{4.20}$$

[1] Note that deleting the values of some of the variables can be seen as destroying part of the incumbent solution, while finding new values for these variables can be regarded as repairing the partially destroyed solution.

where $\overline{S}$ is the index set of the variables that have value one in the incumbent solution $\overline{x}$, that is

$$\overline{S} = \{ j \in \{1,\ldots,n\} \mid \overline{x}_j = 1 \}. \tag{4.21}$$

In the case of binary variables, $\Delta(x,\overline{x})$ resembles the Hamming distance, and thus the neighborhood induced by the local branching constraint corresponds to the classic k-opt neighborhood. Note that parameter k, whose choice is critical, controls the size of the neighborhood. Ways for dynamically adapting k—for example, as in variable neighborhood search [128]—are therefore common. Fischetti and Lodi [98] describe how the concept of local branching constraints can be generalized to non-binary integer variables. However, the major advantage of local branching constraints—namely, that no variables must be explicitly selected for fixing—also comes with a downside: Local branching constraints are dense, in a sense that they involve all binary variables, and their inclusion typically increases the complexity of the MIP model significantly. In particular, including reverse local branching constraints to exclude already searched neighborhoods from consideration in further iterations has not turned out to be fruitful [72].

The local branching approach as outlined above is generally applicable to any problem that can be expressed as (binary) MIP. In contrast, a specialized problem-specific neighborhood can sometimes be defined. For example, Archetti et al. [9] consider the selective arc routing problem with penalties, which is a generalization of the directed rural postman problem in which a minimum cost cycle traversing a subset of arcs at least once is sought; costs arise for unvisited arcs. After performing tabu search, a large neighborhood is defined based on the solutions visited by tabu search: First a minimal tour containing a set of "good" arcs that are likely to be contained in an optimal solution—that is, those arcs that appeared in most of the solutions produced by tabu search—is built. The large neighborhood of this minimal tour consists of all possible extensions by means of sequences of so-called *questionable* arcs. Due to the relatively high running time of the MIP-solver, the large neighborhood search is not iterated here, but only applied once as a final refinement phase. A similar methodology was proposed by De Franceschi et al. [74] for the directed capacitated vehicle routing problem. Further special MIP-based neighborhoods have, for example, been described by Öncan et al. [210] for partitioning problems, Ropke and Pisinger [258] for pickup and delivery problems with time windows, Pirkwieser and Raidl [227] for a periodic location-routing problem, and by Jaśkowski et al. [144] for a machine reassignment problem.

So far we have only dealt with LNS approaches based on the use of MIP solvers. However, the search for good solutions in large neighborhoods is obviously not restricted to the use of MIP solvers. In fact, the related literature contains examples in which large neighborhoods are explored, for example, by dynamic programming (see Section 1.4 for an introduction to dynamic programming). This is because, in some cases, DP can make it possible to completely explore a neighborhood of exponential size in polynomial time and space. An example for such a LNS application is *iterated dynasearch*. An application of iterated dynasearch in the context of the *Single-Machine Total Weighted Tardiness Scheduling Problem* (SMTWTSP) is presented in [62]. The SMTWTSP can be defined as the problem of finding the processing order of n jobs on one machine such that the total tardiness is minimized. More formally, for each job $i \in \{1,\ldots,n\}$ is given a processing time p_j, a positive weight w_j and a due date d_j. Jobs are available at time zero and must be processed one at a time without interruption. Given a valid solution—that is, any permutation of the n jobs—a completion time C_j can be computed for each of the jobs, along with its tardiness $T_j = \max\{\mathrm{C_j} - \mathrm{d_j}, 0\}$. The function to be minimized is $\sum_{j=1}^{n} w_j T_j$. A natural neighborhood of a permutation is obtained by one or more *swaps*. Each swap consists in exchanging two jobs. In the case of allowing exactly one swap, the resulting neighborhood is called the 2-exchange neighborhood. In general, the k-exchange neighborhood, obtained by sequences of swaps involving k objects, has a size of order $O(n^k)$. Therefore, for efficiency concerns, usually only the cases of $k \in \{2,3\}$ are considered. The *dynasearch swap* neighborhood of a job sequence $\sigma = (\sigma(1),\ldots,\sigma(n))$ is composed of all the permutations of σ that can be generated by a series of *independent* swaps. Two swap moves $\{i,j\}$ and $\{k,l\}$ are independent if $\max\{i,j\} < \min\{k,l\}$ or $\min\{i,j\} > \max\{k,l\}$. Note that this neighborhood has size $2^{n-1} - 1$. However, the independence of moves makes it possible to define a recursive enumeration algorithm based on DP such that the resulting exploration is polynomial in time and space. Further contributions to dynasearch and related techniques can be found, for example, in [122, 5, 255]. Other examples of DP-based LNS methods include that for the longest common subsequence problem from [84], that for vehicle routing problems with profits [288], and the *corridor method* for the blocks relocation problems [51].

Finally, LNS based on constraint programming (see Section 1.6 for an introduction to constraint programming) was historically one of the first applications of LNS mentioned in the literature. CP-based LNS denotes a family of problem solving techniques in which local search uses CP for exploring large neighborhoods. This form of LNS was first proposed by Shaw in [268] and similar ideas were presented in [7, 221, 222]. Since its proposal, CP-based LNS was further improved in [269], in which a more efficient mechanism for combining local search and CP was described. Moreover, in [220] LNS is enhanced by a method for automatically generating neighborhoods by means of constraint propagation. Recent approaches of CP-based LNS include the application to the post enrolment-based course timetabling problem [50], where CP-based LNS is applied both as a stand-alone solver and as an intensification procedure for the final stages of simulated annealing.

Chapter 5
Making Use of a Parallel, Non-independent, Construction of Solutions Within Metaheuristics

Most metaheuristics are either based on neighborhood search or the construction of solutions. Examples of the latter include ant colony optimization (ACO) and greedy randomized adaptive search procedures (GRASP). These techniques construct solutions probabilistically by sampling a probability distribution over the search space. Solution constructions are sequential and independent from each other. Moreover, greedy information is used in a probabilistic way as the only source of knowledge on the tackled problem (instance). Note that metaheuristics that are based on solution construction can also be interpreted as incomplete tree search algorithms. In contrast to ACO and GRASP, more classic heuristic tree search methods such as the branch-and-bound derivative known as *beam search*, for example, make use of two different types of problem knowledge. In addition to greedy information, they exploit bounding information in order to prune branches of the search tree and to differentiate between partial solutions on the same level of the search tree. Moreover, these methods can be seen as generating solutions in a (pseudo-)parallel, non-independent way. The general idea of this chapter is to improve metaheuristics based on solution construction by incorporating bounding information and the parallel and non-independent construction of solutions.

The outline of this chapter is as follows. After a more detailed description of the general idea sketched above, a problem-independent algorithmic framework for optimization labeled BEAM-ACO will be described making use of this idea. The BEAM-ACO framework works roughly as follows. At each iteration, a probabilistic beam search algorithm is applied which is based both on greedy and pheromone information as well as on bounding information. The constructed solution is used for updating the pheromone values. In other words, the algorithmic framework of BEAM-ACO is that of ACO. However, instead of performing a number of sequential and independent solution constructions per iteration, a probabilistic beam search algorithm is applied. The *Multidimensional Knapsack Problem* (MKP) is chosen as a test case in order to demonstrate the application of BEAM-ACO. Obtained results show that the proposed algorithm is superior to its pure components, (1) the application of beam search, and (2) the application of standard ant colony optimization.

5.1 General Idea

Most (meta)heuristics for combinatorial optimization are based on one of the following two principles: neighborhood search or solution construction. In neighborhood search we are given a neighborhood function that assigns to each candidate solution a so-called neighborhood, which is a subset of the search space. Heuristic methods based on neighborhood search follow a trajectory in the directed graph G whose node set is the search space. A node v is connected to a node w by an arc if and only if w is in the neighborhood of v. The trajectories in G may be deterministic as in the case of a standard tabu search (see Section 1.2.5) or they might result from a random process as in the case of simulated annealing (see Section 1.2.4).

On the contrary, optimization techniques based on solution construction explore the search space in the form of a search tree which is defined by the solution construction mechanism. Following a path from the root node to a leaf corresponds to the process of constructing a candidate solution. Inner nodes of the tree can be seen as partial solutions. The process of moving from an inner node to one of its child nodes is called a *construction step* or the *extension of a partial solution*. A prominent example of deterministic constructive algorithms are greedy heuristics. They make use of a weighting function that associates weights to a child node of each inner node of the search tree. At each construction step a child node with the highest weight—that is, a most desirable child node—is chosen. Metaheuristics such as ACO (see Section 1.2.7 for an introduction) or GRASP (see Section 1.2.1) employ sequential, independent, probabilistic—that is, randomized—solution constructions. For each inner node of the tree and each child node is given the probability of performing the corresponding construction step. These probabilities, which may depend on greedy functions and possibly the search history of the algorithm, define a probability distribution over the search space. Henceforth, we refer to the problem knowledge used by ACO and GRASP as the *primary problem knowledge*.

In contrast to the above-mentioned metaheuristics based on solution construction, more traditional tree search methods such as branch-and-bound (see also Section 1.3) and incomplete branch-and-bound derivatives such as beam search [261] incorporate—in addition to making use of the primary problem knowledge—the following features:

1. Dual bounds, i.e., lower bounds in case of minimization and upper bounds in case of maximization, are used for evaluating partial solutions; sometimes also for choosing among different partial solutions, or discarding partial solutions together with the whole portion of the search tree rooted in those partial solutions. Henceforth we will refer to this type of knowledge as the *complementary problem knowledge*.
2. The extension of partial solutions may be done in more than one way at the same time. In beam search, the number of nodes that can be selected at each search tree

level is usually limited by a parameter, resulting in *parallel*[1] and *non-independent* solution constructions.

The general idea of this chapter is to incorporate these two features into metaheuristics based on the construction of solutions.

For the remainder of this chapter remember that, given a problem instance $\mathscr{I}$ to a generic problem $\mathscr{P}$, set C represents the *complete set of solution components*, that is, the set of all possible components of which solutions to the problem instance are composed. In the context of this chapter, a valid solution S to $\mathscr{I}$ is represented as a subset of the solution components C, that is, $S \subseteq C$. Moreover, a partial solution S_{partial} is an incomplete solution which can be extended to, at least, one valid solution, and is, therefore, also a subset of C.

5.1.1 *Beam-ACO: Combining Ant Colony Optimization with Beam Search*

A specific example of a metaheuristic that makes use of the complementary problem knowledge and parallel non-independent solution constructions is BEAM-ACO [25], which is an algorithm that was first proposed in 2005 in the context of the open shop scheduling problem. The two basic ingredients of BEAM-ACO are ant colony optimization—which was described in general terms in Section 1.2.7—and the incomplete branch-and-bound derivate *beam search* [261]. The literature offers diverse variants of beam search. In the following we introduce the variant that is used for the application example of this chapter.

Deterministic beam search starts with the empty partial solution $S_{\text{partial}} = \emptyset$. Then, it uses a limited form of breadth-first search in order to traverse the search tree. At each level of the search tree, the current set B of partial solutions—also known as the *beam*—are expanded, each one in at most k_{ext} possible ways. This is done on the basis of a greedy function. The resulting partial solutions are stored in a set B'. Afterwards, set B' is reduced to at most b_{width}—also called the *beam width*—partial solutions. For this purpose, a lower bound value (in the case of minimization problems), respectively an upper bound value (in the case of maximization problems), is computed for each partial solution in B'. Only the b_{width} best partial solutions from B' with respect to the bounding information are transfered to the beam B of the next step of the algorithm. The algorithm stops once B is empty. This whole approach is shown in Algorithm 22. In the context of this algorithm, $g(c \mid S_{\text{partial}})$ denotes the greedy weight of extending partial solution S_{partial} with solution component $c \in C$. Furthermore, given a partial solution S_{partial}, set $Ext(S_{\text{partial}}) \subseteq C$ denotes the set of solution components that may be used in order to extend S_{partial}, resulting again in a

[1] We generally say here that solutions are constructed *in parallel*, although technically this construction does not necessarily happen in a strictly parallel sense but will usually be done in an intertwined, pseudo-parallel way, in particular when only a sequential machine is available.

Algorithm 22 Beam Search (for maximization)

1: **given:** problem instance $\mathcal{I}$
2: $S_{\text{partial}} \leftarrow \emptyset$
3: $B \leftarrow \{S_{\text{partial}}\}$
4: $S_{\text{bsf}} \leftarrow$ NULL
5: **while** B is not empty **do**
6: $B' \leftarrow \emptyset$
7: **for** $S_{\text{partial}} \in B$ **do**
8: **if** S_{partial} is a complete solution **then**
9: **if** $S_{\text{bsf}} =$ NULL **or** $f(S_{\text{partial}}) > f(S_{\text{bsf}})$ **then** $S_{\text{bsf}} \leftarrow S_{\text{partial}}$
10: **else**
11: $E \leftarrow Ext(S_{\text{partial}})$
12: cnt $\leftarrow 0$
13: **while** E not empty **and** cnt $< k_{\text{ext}}$ **do**
14: $c^* \leftarrow \text{argmax}\,\{g(c \mid S_{\text{partial}}) \mid c \in E\}$
15: Produce (partial) solution S^*_{partial} by extending S_{partial} with c^*
16: **if** $\text{UB}(S^*_{\text{partial}}) > f(S_{\text{bsf}})$ **then**
17: $E \leftarrow E \setminus \{c^*\}$
18: $B' \leftarrow B' \cup \{S^*_{\text{partial}}\}$
19: cnt $\leftarrow$ cnt $+ 1$
20: **end if**
21: **end while**
22: **end if**
23: **end for**
24: determine B as the best $\min\{|B'|, b_{\text{width}}\}$ partial solutions from B' w.r.t. their UB() values
25: **end while**
26: **return** S_{bsf}

valid (partial) solution. Finally, $\text{UB}(S_{\text{partial}})$ denotes the upper bound value of S_{partial}. Note that an upper bound is used because the application example of this chapter concerns a maximization problem.

Beam search has been applied as a standalone technique in quite a number of research papers during recent decades. Examples include scheduling problems [211, 264, 108], berth allocation [297], assembly line balancing [88], and circular packing [3]. In contrast, within BEAM-ACO beam search is used in an adaptive, probabilistic way in order to replace the independent construction of n_{a} solutions per iteration—that is, at each iteration BEAM-ACO employs exactly one probabilistic beam search algorithm in order to construct b_{width} solutions non-independently in parallel. For this purpose, line 14 of Algorithm 22 is replaced by the choice of a solution component $c^* \in E$ in a probabilistic way, not only depending on the greedy information $g(c \mid S_{\text{partial}})$ but also on pheromone information. In the remainder of this chapter we demonstrate the application of BEAM-ACO in the context of a specific example.

5.2 Application to the Multidimensional Knapsack Problem

Knapsack problems are well-studied combinatorial optimization problems. To illustrate the application of BEAM-ACO we chose the so-called *Multidimensional Knapsack Problem* (MKP). In contrast to the basic (one-dimensional) knapsack problem as introduced in Section 1.4 for illustrating dynamic programming, the MKP is strongly NP-hard. In general, dynamic programming does not work here so efficiently anymore.

The MKP can technically be stated as follows. Given are a set $I = \{1,\ldots,n\}$ of items and a set $K = \{1,\ldots,m\}$ of different resources. Each resource $k \in K$ is available in a certain quantity (*capacity*) $c_k > 0$, and each item $i \in I$ requires from each resource $k \in K$ a given amount $r_{i,k} > 0$ (*resource consumption*). Moreover, each item $i \in I$ has associated a profit $p_i > 0$.

A feasible solution to the MKP is a selection (subset) of items $S \subseteq I$ such that for each resource k, the total consumption over all selected items $\sum_{i \in S} r_{i,k}$ does not exceed the resource's capacity c_k. As a simple example, consider packing a real knapsack and besides a limit on the total weight one can carry, one also has to take care not to exceed the knapsack's available volume; thus we deal with $m = 2$ resources—weight and volume—in this case. As in the basic knapsack problem, the MKP's objective is to find a feasible item selection S of maximum total profit $\sum_{i \in S} p_i$.

Formally, the MKP can be stated by the following ILP, which uses a binary incidence vector $x = (x_1,\ldots,x_n) \in \{0,1\}^n$ for indicating the selected items, i.e., $i \in S \leftrightarrow x_i = 1$, for $i = 1,\ldots,n$.

$$\text{maximize} \quad \sum_{i=1}^{n} p_i \cdot x_i \tag{5.1}$$

$$\text{subject to} \quad \sum_{i=1}^{n} r_{i,k} \cdot x_i \leq c_k \qquad \forall k = 1,\ldots,m \tag{5.2}$$

$$x_i \in \{0,1\} \qquad \forall i = 1,\ldots,n \tag{5.3}$$

Inequalities (5.2) limit the total consumption for each resource and are called *knapsack constraints*.

The MKP has been quite popular as a benchmark case for new algorithmic proposals during recent decades. Exact methods proposed to solve the MKP include dynamic programming [16] (which is typically only effective for small m or restricted resource consumption values), hybrid dynamic programming [43] and branch-and-bound [289, 174]. Metaheuristics applied to the MKP include evolutionary algorithms [57, 298], simulated annealing [173], tabu search [125], ant colony optimization [163], particle swarm optimization [53], and harmony search [164]. The remainder of this section describes a simple greedy heuristic, a beam search approach, a pure ACO metaheuristic, and the application of BEAM-ACO to the MKP problem. Note that the purpose of this application example is to demonstrate BEAM-ACO, rather than producing state-of-the-art results.

Algorithm 23 Greedy Heuristic for the MKP

1: **given:** MKP instance
2: $S \leftarrow \emptyset$
3: **for** $i \leftarrow 1,\ldots,n$ **do**
4: **if** $\left(\sum_{j \in S} r_{j,k}\right) + r_{i,k} \leq c_k,\ \forall k = 1,\ldots,m$ **then**
5: $S \leftarrow S \cup \{i\}$
6: **end if**
7: **end for**
8: **return** S

5.2.1 Greedy Heuristic

The following greedy heuristic was considered as the basis of the beam search and the ACO approach to be described below. First, we henceforth assume the items in I are ordered w.r.t. the following *utility values* in a non-increasing way:

$$u_i \leftarrow \frac{p_i}{\sum_{k=1}^{m} r_{i,k}/c_k} \qquad i = 1,\ldots,n. \tag{5.4}$$

That is, the items in I are ordered such that $u_1 \geq u_2 \geq \ldots \geq u_n$. The values u_i are used as a static greedy weighting function in the greedy algorithm shown in Algorithm 23. This heuristic simply adds items in the order determined by the utility values to an initially empty partial solution S until no further item fits w.r.t. the remaining resource capacities.

5.2.2 Beam Search for the MKP

The beam search algorithm for the MKP is provided in Algorithm 24. In the context of this algorithm, given a partial solution S_{partial}, we denote by $l(S_{\text{partial}})$ the last item that was added to S_{partial} (see line 11 of Algorithm 24). Moreover, the upper bound value $\text{UB}(S_{\text{partial}})$ of a partial solution S_{partial}—as needed in line 25—is obtained by solving the linear programming relaxation of the ILP model (5.1)–(5.3) with CPLEX. Apart from relaxing the variable domains (5.3) to $0 \leq x_i \leq 1$, $i = 1,\ldots,n$, the values of those variables that correspond to items in S_{partial} are fixed to one.

Let us, for a moment, consider the requirements on the bounding information used within beam search. In breadth-first branch-and-bound, generally, it is important that bounds are usually reasonably tight. This means that there should not be a large gap between the value of the best solution that can be generated on the basis of a partial solution S_{partial} and the value of the bounding information for S_{partial}. If the bounding information is tight, breadth-first branch-and-bound is able to cut away large parts of the search tree, implying that the algorithm will terminate much faster than complete enumeration. In contrast, beam search is less dependent on the tightness of the bounding information. In fact, it is much more important that the

Algorithm 24 Beam Search for the MKP

```
1: given: MKP instance, k_ext, and b_width
2: S_partial ← ∅
3: B ← {S_partial}
4: S_bsf ← NULL
5: while B is not empty do
6:     B' ← ∅
7:     for S_partial ∈ B do
8:         if S_partial is not extensible by any item in I \ S_partial then
9:             if S_bsf = NULL or f(S_partial) > f(S_bsf) then S_bsf ← S_partial
10:        else
11:            i ← l(S_partial) + 1
12:            cnt ← 0
13:            while i ≤ n and cnt < k_ext do
14:                if (Σ_{j∈S_partial} r_{j,k}) + r_{i,k} ≤ c_k, ∀k = 1,...,m then
15:                    S'_partial ← S_partial ∪ {i}
16:                    if UB(S'_partial) > f(S_bsf) then
17:                        B' ← B' ∪ {S'_partial}
18:                        cnt ← cnt + 1
19:                    end if
20:                end if
21:                i ← i + 1
22:            end while
23:        end if
24:    end for
25:    determine B as the best min{|B'|, b_width} partial solutions from B' w.r.t. their UB() values
26: end while
27: return S_bsf
```

bounding information allows discriminating well between partial solutions from the same level of the search tree, that is, between partial solutions containing the same number of solution components. By discriminating well between different partial solutions we refer to the ability to correctly predict that a partial solution S'_{partial} may lead to a better complete solution than another partial solution S''_{partial}. An experimental evaluation of the discriminatory power of the LP relaxation in the case of the MKP is presented in Section 5.2.5.1.

5.2.3 A Pure ACO Approach for the MKP

The ACO variant that we chose for the MKP is known as a MAX-MIN Ant System (MMAS) implemented in the Hyper-Cube Framework (HCF) [29]. In the following, we provide a description of this pure ACO approach, henceforth simply labeled ACO.

As in the case of GREEDY and BEAM SEARCH, a solution S in the context of ACO is simply a (sub-)set of the set I of items, that is, $S \subseteq I$. The pheromone model

Algorithm 25 ACO for the MKP

1: **given:** MKP instance, d_{rate}, l_{size}
2: $S_{\text{bsf}} \leftarrow \emptyset$, $S_{\text{rb}} \leftarrow \emptyset$, $cf \leftarrow 0$, $bs\_update \leftarrow$ **false**
3: $\tau_i \leftarrow 0.5$ for all $\tau_i \in \mathscr{T}$
4: **while** CPU time limit not reached **do**
5:    **for** cnt $\leftarrow 1,\ldots,n_a$ **do**
6:       $S_{\text{cnt}} \leftarrow$ ConstructSolution() {see Algorithm 26}
7:    **end for**
8:    $S_{\text{ib}} \leftarrow \text{argmin}\,\{f(S_{\text{cnt}}) \mid \text{cnt} = 1,\ldots,n_a\}$
9:    **if** $f(S_{\text{ib}}) > f(S_{\text{rb}})$ **then** $S_{\text{rb}} \leftarrow S_{\text{ib}}$
10:    **if** $f(S_{\text{ib}}) > f(S_{\text{bsf}})$ **then** $S_{\text{bsf}} \leftarrow S_{\text{ib}}$
11:    ApplyPheromoneUpdate(cf,bs_update,$\mathscr{T}$,S_{ib},S_{rb},S_{bsf})
12:    $cf \leftarrow$ ComputeConvergenceFactor($\mathscr{T}$)
13:    **if** $cf > 0.99$ **then**
14:       **if** $bs\_update =$ **true then**
15:          $\tau_i \leftarrow 0.5$ for all $\tau_i \in \mathscr{T}$
16:          $S_{\text{rb}} \leftarrow$ NULL
17:          $bs\_update \leftarrow$ **false**
18:       **else**
19:          $bs\_update \leftarrow$ **true**
20:       **end if**
21:    **end if**
22: **end while**
23: **return** S_{bsf}, the best solution

$\mathscr{T}$ consists of a pheromone value $\tau_i \geq 0$ for each item $i \in I$. The pseudo-code of ACO is presented in Algorithm 25. First, n_a solutions are probabilistically generated, based on pheromone and greedy information. Second, the pheromone values are modified using (at most) three solutions: (1) the iteration-best solution S_{ib}, (2) the restart-best solution S_{rb}, and (3) the best-so-far solution S_{bsf}. The general aim of the pheromone update is to focus the search on areas of the search space containing high-quality solutions. Moreover, the algorithm performs restarts—that is, re-initializations of the pheromone values—upon convergence. More specifically, restarts are controlled by the so-called convergence factor (cf) and a Boolean control variable called *bs_update*. A detailed description of all algorithmic components is provided in the following.

ConstructSolution(): A solution S is generated in a probabilistic way, based on the solution construction mechanism of GREEDY as outlined in Algorithm 23. For each solution construction the items from I are re-ordered in a non-increasing way with respect to $u_i \cdot \tau_i$; for the definition of the utility values u_i see Equation (5.4). The resulting order of the items is stored in a permutation π, where $\pi(j)$ denotes the item at position $j = 1,\ldots,n$, and $\pi^{-1}(i)$ denotes the position of item $i \in I$ in permutation π. In contrast to GREEDY, in which the first item after position $l(S)$ that fits w.r.t. all remaining resource capacities is added to solution S, in the context of ACO the choice of the next item works as follows. At each iteration, the first up to l_{size} items starting from position $\pi^{-1}(l(S))+1$ in π that fit w.r.t. all resources are collected in a set L. Note that l_{size} is an important algorithm parameter. Then,

Algorithm 26 Function ConstructSolution() of Algorithm 25

1: **given:** MKP instance, pheromone model $\mathscr{T}$
2: $\pi \leftarrow$ order of the items in I according to non-increasing values of $u_i \cdot \tau_i$
3: $S \leftarrow \emptyset$
4: $L \leftarrow$ First up to l_{size} items starting from position 1 in π that fit into S without constraint violations
5: **while** $L \neq \emptyset$ **do**
6: choose $v \in [0,1)$ uniformly at random
7: **if** $v \leq d_{\text{rate}}$ **then**
8: $i^* \leftarrow \text{argmin}\{\pi^{-1}(i) \mid i \in L\}$
9: **else**
10: $i^* \leftarrow$ choose uniformly at random from L
11: **end if**
12: $S \leftarrow S \cup \{i^*\}$
13: $L \leftarrow$ first up to l_{size} items starting from position $\pi^{-1}(l(S))+1$ in π that fit into S without constraint violation
14: **end while**
15: **return** S

a random number $v \in [0,1)$ is chosen uniformly. In case $v \leq d_{\text{rate}}$, the item $i^* \leftarrow \text{argmin}\{\pi^{-1}(i) \mid i \in L\}$ is chosen and added to S. Just like l_{size}, the determinism rate d_{rate} is an input parameter of the algorithm for which a well-working value must be found. Otherwise—that is, in case $v > d_{\text{rate}}$—an item i^* from L is chosen uniformly at random.

ApplyPheromoneUpdate(cf,bs_update,$\mathscr{T}$,S_{ib},S_{rb},S_{bsf}): The pheromone update is performed in the same way as in all MMAS algorithms implemented in the HCF. The three solutions S_{ib}, S_{rb}, and S_{bsf} (as described at the beginning of this section) are used for the pheromone update. The influence of these three solutions on the pheromone update is determined by the current value of the convergence factor cf. Each pheromone value $\tau_i \in \mathscr{T}$ is updated by

$$\tau_i \leftarrow \tau_i + \rho \cdot (\xi_i - \tau_i), \tag{5.5}$$

with

$$\xi_i \leftarrow \kappa_{ib} \cdot \Delta(S_{\text{ib}}, i) + \kappa_{rb} \cdot \Delta(S_{\text{rb}}, i) + \kappa_{bsf} \cdot \Delta(S_{\text{bsf}}, i), \tag{5.6}$$

where κ_{ib} is the weight of solution S_{ib}, κ_{rb} that of solution S_{rb}, and κ_{bsf} that of solution S_{bsf}. Moreover, $\Delta(S,i) = 1$ if and only if item $i \in S$ and 0 otherwise. The three weights must be chosen such that $\kappa_{ib} + \kappa_{rb} + \kappa_{bsf} = 1$ holds. After the application of Equation (5.5), pheromone values exceeding $\tau_{\max} = 0.99$ are reset to $\tau_{\max}$, and pheromone values that have fallen below $\tau_{\min} = 0.01$ are reset to $\tau_{\min}$. This prevents the algorithm from reaching a state of convergence. Finally, note that the exact values of the weights depend on the convergence factor cf and on the value of the Boolean control variable bs_update as outlined in Table 5.1.

ComputeConvergenceFactor($\mathscr{T}$): The convergence factor cf is computed on the basis of the pheromone values according to [29]

Table 5.1 Setting of κ_{ib}, κ_{rb}, and κ_{bsf} in dependence of the convergence factor *cf* and the Boolean control variable *bs_update*

	bs_update = FALSE				*bs_update*
	$cf < 0.4$	$cf \in [0.4, 0.6)$	$cf \in [0.6, 0.8)$	$cf \geq 0.8$	= TRUE
κ_{ib}	1	2/3	1/3	0	0
κ_{rb}	0	1/3	2/3	1	0
κ_{bs}	0	0	0	0	1

$$cf \leftarrow 2 \cdot \left(\left(\frac{\sum_{\tau_i \in \mathcal{T}} \max\{\tau_{\max} - \tau_i, \tau_i - \tau_{\min}\}}{|\mathcal{T}| \cdot (\tau_{\max} - \tau_{\min})} \right) - 0.5 \right)$$

This results in $cf = 0$ when all pheromone values are 0.5. On the contrary, when all pheromone values have either value $\tau_{\min}$ or $\tau_{\max}$, then $cf = 1$. In all other cases, *cf* has a value between 0 and 1. This completes the description of all components of the proposed algorithm.

5.2.4 Beam-ACO for the MKP

As mentioned already at the beginning of this chapter, the algorithmic framework of BEAM-ACO corresponds to that of ACO. The only difference is the following one. Lines 5–7 of Algorithm 25—that is, the independent construction of n_a solutions—is replaced by one execution of a slightly changed version of BEAM SEARCH outlined in Algorithm 24. This change concerns the way in which partial solutions are extended (lines 12–23). In fact, the extension of a partial solution is done in nearly the same way as in ACO. First, a set L of the best possible options for extending the current partial solution is identified, in the same way as in ACO. The size of set L is chosen as $2k_{\text{ext}}$. Then, k_{ext} extensions are chosen from L in the same way as in ACO.

5.2.5 Experimental Evaluation

The proposed application of BEAM-ACO to the MKP was implemented in ANSI C++ using GCC 4.7.3 for compiling the software. Moreover, the LP relaxations used as upper bounds within BEAM SEARCH and BEAM-ACO were solved with IBM ILOG CPLEX 12.1. The experimental evaluation was conducted on a cluster of 32 PCs with Intel Xeon CPU X5660 CPUs of 2 nuclei of 2.8 GHz and 48 GB of RAM. The following algorithms were considered for the comparison:

1. GREEDY: the greedy approach shown in Algorithm 23.
2. BEAM SEARCH: the beam search approach from Section 5.2.2.

3. ACO: the pure ACO approach described in Section 5.2.3.
4. BEAM-ACO: the hybrid approach from Section 5.2.4.

With the aim of testing those algorithms in different scenarios we created the following diverse set of benchmark instances by using the methodology described in [104, 57]. In particular, we generated benchmark instances of $n \in \{50, 100, 150, 200, 250\}$ items. Moreover, $m \in \{5, 10, 15\}$ resources were considered. Finally, the tightness of the instances concerning the knapsack capacities—as expressed by the numerical parameter $\alpha \in (0, 1]$—was chosen from $\alpha \in \{0.25, 0.5, 0.75\}$. Hereby, the lower the value of α, the tighter is a problem instance. Ten random instances were created for each combination of the three above-mentioned parameters. For all instances, the resource requirements $r_{i,j}$ were chosen uniformly at random from $\{1, \ldots, 1000\}$. In total, the generated benchmark set consists of 450 problem instances.

5.2.5.1 Discriminatory Power of the Bounding Information

As already described in Section 5.2.2, discriminating well between different partial solutions is an important requirement for bounding information used in beam search. We say that a lower (respectively, upper) bound discriminates well if, in most cases, the bound correctly predicts that a partial solution S'_{partial} leads to a better complete solution than another partial solution S''_{partial}. In order to test the discriminatory power of the LP relaxation which is used as upper bound in BEAM SEARCH and in BEAM-ACO, we conducted the following experiments. First of all, we generated a set of problem instances small enough in order to explore the whole search tree in a reasonable computation time. We generated 10 random instances in the same way as outlined above for each combination of $n = 20$, $n \in \{5, 10, 15\}$, and $\alpha \in \{0.25, 0.5, 0.75\}$. After generating the complete search tree for each of these instances, we compared for each pair of partial solutions $(S'_{\text{partial}}, S''_{\text{partial}})$ on the same level of each tree their upper bound values $\text{UB}(S'_{\text{partial}})$ and $\text{UB}(S''_{\text{partial}})$ and the values of the best complete solutions that can be built on the basis of these partial solutions, henceforth denoted by $b(S'_{\text{partial}})$ and $b(S''_{\text{partial}})$. We say that the upper bound *does not discriminate well* between two partial solutions S'_{partial} and S''_{partial} if and only if (1) $\text{UB}(S'_{\text{partial}}) < \text{UB}(S''_{\text{partial}})$ and $b(S'_{\text{partial}}) > b(S''_{\text{partial}})$ or if (2) $\text{UB}(S'_{\text{partial}}) > \text{UB}(S''_{\text{partial}})$ and $b(S'_{\text{partial}}) < b(S''_{\text{partial}})$. In all other cases we say that the upper bound discriminates well. In Table 5.2 we present the percentage of cases in which the upper bound discriminates well, averaged over 10 problem instances. A *case* refers hereby to a comparison of two partial solutions on the same level of the search tree. It can be observed first that—in all cases—the percentage of correct decisions is above 80.0%. Second, with decreasing instance tightness the quality of the upper bound, in terms of its discriminatory power, increases. Third, it can be observed that the number of resources m does not seem to have an influence on the discriminatory power of the upper bound. Given these results, it can be expected that BEAM SEARCH and BEAM-ACO perform well.

Table 5.2 Percentage of correct decisions taken on the basis of the upper bound information. Instances are characterized by the number of resources m and the tightness α

	$m=5$	$m=10$	$m=15$
$\alpha=0.25$	85.7%	83.8%	84.5%
$\alpha=0.50$	90.6%	89.2%	90.4%
$\alpha=0.75$	96.0%	93.8%	95.3%

Table 5.3 Parameter settings produced by irace for the three different instance tightness levels

	n_a	l_{size}
$\alpha=0.25$	15	5
$\alpha=0.50$	15	5
$\alpha=0.75$	25	5

(a) Tuning results for ACO

	d_{rate}	k_{ext}	b_{width}
$\alpha=0.25$	0.6	5	15
$\alpha=0.50$	0.9	5	25
$\alpha=0.75$	0.6	5	20

(b) Tuning results for BEAM-ACO

5.2.5.2 Tuning of ACO and BEAM-ACO

Both ACO and BEAM-ACO have several parameters for which well-working values must be found. First, less critical parameters were determined after initial tests. The learning rate l_{rate} of both algorithms was set to 0.1, and it has been observed that a single fixed value for the determinism rate d_{rate} does not work well in the context of ACO. Therefore, after fixing a value for n_a (the number of independent solution constructions per iteration), the determinism rate is chosen as follows. For the j-th solution construction of an iteration, with $j=1,\ldots,n_a$, d_{rate} is set to $(2j-1)/2n_a$. As an example, in case of $n_a=10$, the values for d_{rate} for the 10 solution constructions per iteration are $\{0.05,0.15,\ldots,0.95\}$. In this way, at each iteration solutions are constructed with different values for d_{rate}, ranging from very low to very high rates of determinism.

The remaining parameters of both algorithms were subject to systematic tuning. In the case of ACO, this concerns (1) parameter n_a and (2) parameter l_{size}. In the case of BEAM-ACO, this concerns (1) the determinism rate d_{rate}, (2) the number of allowed extensions per partial solution (k_{ext}), and (3) the beam width b_{width}. The following domains were considered: $n_a, l_{size}, k_{ext}, b_{width} \in \{5,10,15,20,25\}$, and $d_{rate} \in \{0.0,0.3,0.6,0.9\}$.

The automatic configuration tool irace [179] was used for the fine-tuning of the parameters. In fact, irace was applied to tune ACO as well as BEAM-ACO separately for each instance tightness in $\{0.25,0.50,0.75\}$. In total, 15 *training instances* for each of the three different instance tightness levels were generated for tuning: one random instance for each combination of $n \in \{50,100,150,200,250\}$ and $m \in \{5,10,15\}$. The tuning process for each tightness level was given a budget of 1000 runs of ACO, respectively BEAM-ACO. For each run of the two algorithms a computation time limit of $2n$ CPU seconds was imposed.

The three applications of irace for each of the two algorithms produced the three sets of parameter configurations as shown in Tables 5.3a and 5.3b.

Table 5.4 Numerical results for the MKP instances with $m = 5$ resources

n	**tightness**	GREEDY		BEAM SEARCH		ACO		BEAM-ACO	
		mean	**time**	**mean**	**time**	**mean**	**time**	**mean**	**time**
	$\alpha = 0.25$	11184.5	0.0	11775.1	0.5	11816.8	0.3	11814.9	7.6
50	$\alpha = 0.50$	18676.6	0.0	20209.2	1.4	20238.6	9.3	20230.8	12.3
	$\alpha = 0.75$	28381.9	0.0	29701.8	1.6	29702.8	5.7	29705.6	12.2
	$\alpha = 0.25$	21292.0	0.0	22777.9	1.6	22865.5	40.7	22851.8	64.6
100	$\alpha = 0.50$	43398.0	0.0	45159.5	4.8	45166.3	79.1	45186.6	21.9
	$\alpha = 0.75$	56366.0	0.0	59089.7	4.6	59070.7	80.2	59142.9	41.6
	$\alpha = 0.25$	34772.3	0.0	36978.5	3.9	37055.1	123.2	37071.1	46.5
150	$\alpha = 0.50$	62906.3	0.0	66028.8	9.9	66005.1	138.1	66135.1	70.0
	$\alpha = 0.75$	89106.9	0.0	91541.7	10.0	91444.6	106.8	91616.5	101.2
	$\alpha = 0.25$	46618.7	0.0	48828.7	7.5	48840.4	123.5	48899.5	93.7
200	$\alpha = 0.50$	85783.7	0.0	87835.6	19.6	87798.9	129.7	87909.8	109.9
	$\alpha = 0.75$	118787.1	0.0	121999.8	18.3	121873.8	215.5	122074.6	211.1
	$\alpha = 0.25$	58356.2	0.0	60585.8	12.5	60570.9	158.7	60645.3	204.9
250	$\alpha = 0.50$	105528.4	0.0	109331.8	31.4	109266.7	314.6	109424.2	190.4
	$\alpha = 0.75$	147876.0	0.0	152496.8	29.6	152133.4	213.5	152548.7	337.6

5.2.5.3 Results

Numerical results are shown in Table 5.4 for the instances with $m = 5$ resources, in Table 5.5 for instances with $m = 10$ resources, and in Table 5.6 for instances with $m = 15$ resources. The results are presented in these tables in terms of averages (means) over 10 random instances of the same characteristics. Each algorithm included in the comparison was applied exactly once to each problem instance, with a computation time limit of $2n$ CPU seconds per problem instance. Note that the results of BEAM SEARCH were obtained with the same values for k_{ext} and b_{width} as BEAM-ACO.

The structure of the three tables is as follows. The first column provides the number of items, whereas the second column indicates the tightness level of the corresponding instances. The results of GREEDY, BEAM SEARCH, ACO and BEAM-ACO are presented in two columns each, in terms of the obtained average solution quality (column **mean**) and the average computation time in seconds needed to produce these results (column **time**). To conclude, the best result of each table row is marked by a gray background.

Apart from the numerical results provided in the form of tables, Figure 5.1 shows the improvement of BEAM-ACO over ACO for all instances and Figure 5.2 shows the improvement of BEAM-ACO over BEAM SEARCH for all instances. The notation X-Y on the x-axis of these graphics has the following meaning. X and Y take values from $\{S, M, L\}$. While X refers to the number of resources, Y refers to the tightness level. In case of X, S refers to 5, M to 10, and L to 15, while in the case of Y, S refers to 0.25, M to 0.50, and L to 0.75.

The following observations can be made:

Table 5.5 Numerical results for the MKP instances with $m = 10$ resources

n	**tightness**	GREEDY		BEAM SEARCH		ACO		BEAM-ACO	
		mean	**time**	**mean**	**time**	**mean**	**time**	**mean**	**time**
	$\alpha = 0.25$	10078.2	0.0	10819.6	0.6	10889.3	0.3	10887.5	9.4
50	$\alpha = 0.50$	18813.8	0.0	20350.7	1.7	20414.4	6.0	20402.8	5.3
	$\alpha = 0.75$	26732.6	0.0	28037.5	1.8	28049.2	10.6	28054.9	14.3
	$\alpha = 0.25$	20982.5	0.0	22521.4	2.1	22631.9	44.6	22627.7	37.7
100	$\alpha = 0.50$	38551.6	0.0	41181.0	5.9	41215.1	71.0	41305.0	27.5
	$\alpha = 0.75$	56539.4	0.0	58914.2	6.1	58815.0	44.5	58956.1	49.6
	$\alpha = 0.25$	32724.6	0.0	34486.7	5.3	34646.5	49.0	34735.2	142.6
150	$\alpha = 0.50$	60578.9	0.0	64076.5	14.5	63982.6	146.7	64129.6	77.9
	$\alpha = 0.75$	86527.1	0.0	89724.4	14.2	89599.9	90.6	89784.9	141.3
	$\alpha = 0.25$	42892.3	0.0	46261.7	10.8	46243.6	112.8	46373.7	145.1
200	$\alpha = 0.50$	80692.2	0.0	84404.9	28.5	84173.8	128.5	84542.9	146.2
	$\alpha = 0.75$	113938.0	0.0	118585.1	27.6	118448.9	201.4	118767.9	242.7
	$\alpha = 0.25$	54783.3	0.0	58344.0	19.2	58370.7	187.8	58521.6	167.8
250	$\alpha = 0.50$	104428.4	0.0	108620.6	49.6	108401.2	269.6	108802.2	270.1
	$\alpha = 0.75$	143729.3	0.0	149104.5	45.2	148803.8	245.2	149229.6	331.8

Table 5.6 Numerical results for the MKP instances with $m = 15$ resources

n	**tightness**	GREEDY		BEAM SEARCH		ACO		BEAM-ACO	
		mean	**time**	**mean**	**time**	**mean**	**time**	**mean**	**time**
	$\alpha = 0.25$	9050.4	0.0	10220.1	0.6	10316.6	0.1	10316.6	2.7
50	$\alpha = 0.50$	17584.5	0.0	19209.6	1.9	19246.4	14.0	19243.4	6.2
	$\alpha = 0.75$	26178.8	0.0	27704.5	2.0	27700.3	10.0	27710.0	7.7
	$\alpha = 0.25$	19822.0	0.0	22040.9	2.6	22188.1	68.5	22187.9	51.4
100	$\alpha = 0.50$	39406.9	0.0	41546.1	7.6	41543.5	65.6	41612.2	57.2
	$\alpha = 0.75$	55473.6	0.0	58254.6	7.4	58235.7	55.7	58371.5	63.2
	$\alpha = 0.25$	30645.4	0.0	33150.7	6.7	33261.1	173.9	33297.8	165.1
150	$\alpha = 0.50$	59754.2	0.0	63222.9	19.9	63133.4	157.7	63351.9	116.5
	$\alpha = 0.75$	84868.4	0.0	87898.3	19.3	87822.7	103.8	87997.0	175.9
	$\alpha = 0.25$	42531.5	0.0	45570.7	14.4	45588.3	160.8	45724.7	182.9
200	$\alpha = 0.50$	79044.5	0.0	83859.5	37.7	83616.2	169.6	84005.4	132.9
	$\alpha = 0.75$	112292.6	0.0	116493.7	36.6	116291.8	202.0	116651.8	211.3
	$\alpha = 0.25$	53670.6	0.0	56992.9	26.3	57015.3	281.3	57274.1	197.1
250	$\alpha = 0.50$	102424.5	0.0	107586.1	68.3	107212.5	253.5	107701.0	245.4
	$\alpha = 0.75$	145158.6	0.0	149004.0	63.5	148702.4	154.9	149034.9	399.9

- As expected, ACO, BEAM SEARCH and BEAM-ACO substantially outperform GREEDY. This means that the learning mechanism employed by ACO is beneficial and helps ACO to overcome the partially wrong guidance of the primary problem knowledge used in GREEDY. In the context of BEAM SEARCH and BEAM-ACO this shows that the additional use of the secondary problem knowledge in the form of bounding information is useful. This was to be expected after studying the discriminatory power of the LP relaxation as presented in Section 5.2.5.1.

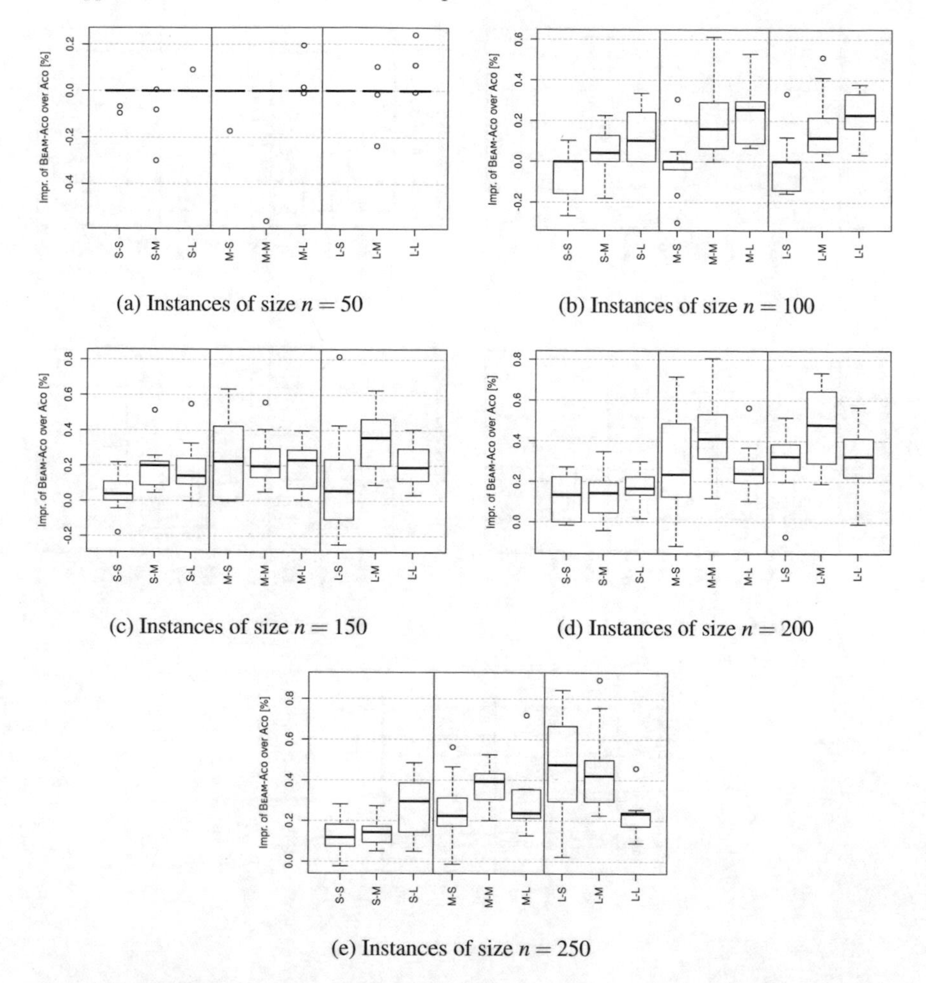

(a) Instances of size $n = 50$

(b) Instances of size $n = 100$

(c) Instances of size $n = 150$

(d) Instances of size $n = 200$

(e) Instances of size $n = 250$

Fig. 5.1 Relative differences between the results of BEAM-ACO and those obtained by ACO (in percent) concerning all instances from the considered benchmark set. Each box shows these differences for the corresponding 10 instances. Note that negative values indicate that ACO obtained a better result than BEAM-ACO

- BEAM-ACO generally outperforms both ACO and BEAM SEARCH. This indicates that the simultaneous use of primary and secondary problem knowledge is useful.
- The relative improvement of BEAM-ACO over BEAM SEARCH is greater for instances with a high tightness level. Interestingly, the opposite is often the case—especially for instances with a low number of resources—in the context of the improvement of BEAM-ACO over ACO; that is, the improvement of BEAM-ACO over ACO is often higher (in percent) for instances with a low tightness level.

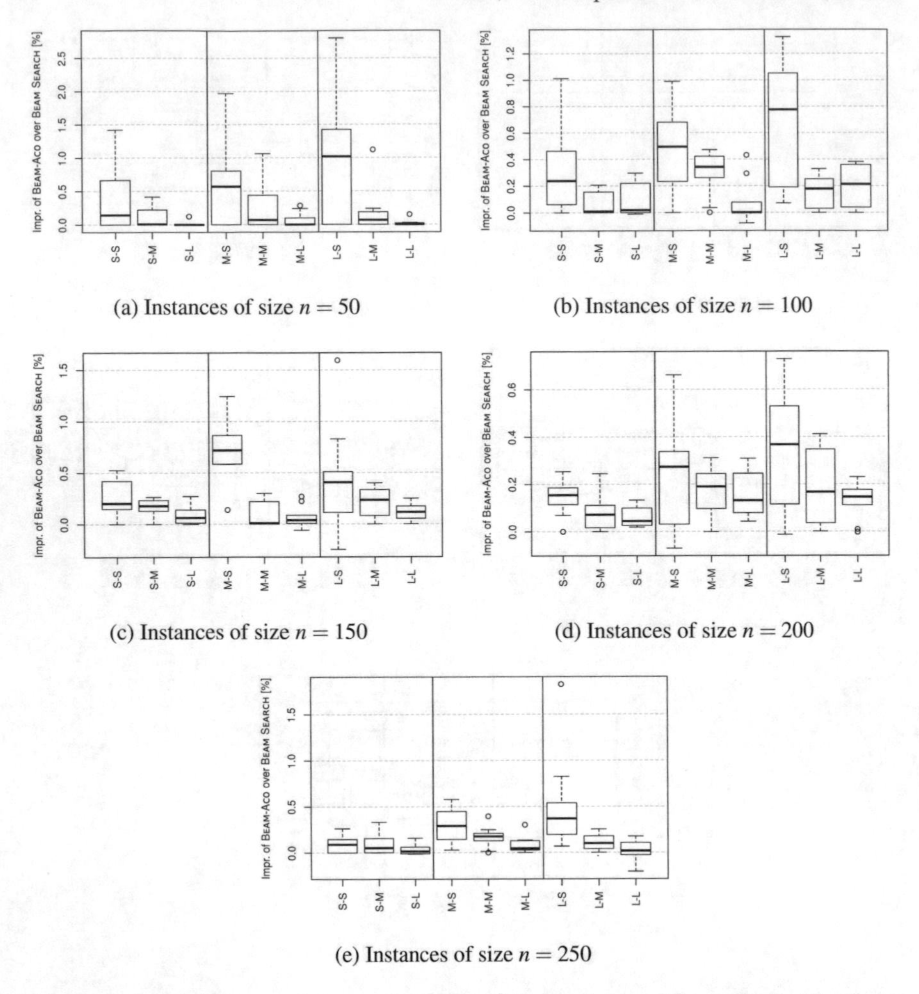

(a) Instances of size $n = 50$

(b) Instances of size $n = 100$

(c) Instances of size $n = 150$

(d) Instances of size $n = 200$

(e) Instances of size $n = 250$

Fig. 5.2 Relative differences between the results of BEAM-ACO and those obtained by BEAM SEARCH (in percent) concerning all instances from the considered benchmark set. Each box shows these differences for the corresponding 10 instances. Note that negative values indicate that BEAM SEARCH obtained a better result than BEAM-ACO

5.3 Other Applications of the Idea of Parallel, Non-independent Solution Construction

The example for the use of parallel, non-independent solution constructions within metaheuristics that was chosen for this chapter—namely Beam-ACO—has been used in the related literature for solving several hard combinatorial optimization problems quite successfully. The first application published in the literature was that to the open shop scheduling problem [25]. Other applications of the standard Beam-ACO algorithm concern diverse variants of the assembly line balancing problem [26, 33, 304], supply chain management [47, 48], weighted vehicle routing [281], and two variants of the longest common subsequence problem [27, 35].

A different Beam-ACO variant concerns the additional hybridization with constraint programming (see Section 1.6 for an introduction to CP). In general, ACO and CP are constructive techniques with complementary strengths: ACO excels by its learning capabilities, while CP is very effective in handling problem constraints. A general framework for combining ACO and CP was proposed in [273]. The additional hybridization with Beam-ACO was done for highly constrained problems in the context of a single machine scheduling problem [285], car sequencing [282], and a shared resource constrained scheduling [283]. It was shown that the hybridization with CP increases the ability of Beam-ACO to find feasible solutions.

Another variant of Beam-ACO was proposed in the context of the traveling salesman problem with time windows (TSPTW) [178, 177, 176]. Due to the lack of appropriate bounding information, the authors replaced the bounding information by *stochastic sampling*. More specifically, given a partial solution, an approximate value for the complementary problem knowledge was obtained by stochastically generating a fixed number of complete solutions on the basis of the partial solution. The value of the best one of these solutions was used as the value of the complementary problem knowledge.

In the context of some applications it has been shown that not all components of a Beam-ACO algorithm may actually be contributing to the algorithm's success. This was the case, for example, in the context of the application to the shortest common super-sequence problem [32]. The resulting algorithm is a probabilistic beam search algorithm, that is, a Beam-ACO algorithm without the learning component based on pheromone values. An improved version of probabilistic beam search for the same problem is presented in [105]. In this work, probabilistic beam search is used both for the purpose of generating initial solutions and for improving solutions. Another probabilistic beam search approach—named stochastic beam search there—was proposed in [297] for a berth allocation problem.

The work on so-called *Approximate and Non-deterministic Tree Search* (ANTS) procedures [184, 185, 186] is partly related to Beam-ACO and probabilistic beam search. In these works, bounding information is used to evaluate the extensions of partial solutions as well as for pruning parts of the search tree. Another way of hybridizing metaheuristics with beam search is presented in [266, 212] in the context of two variants of the open shop scheduling problem. Beam search is used in these works as a decoder for translating individuals from the population of a particle swarm optimization algorithm, respectively a genetic algorithm, into solutions to the tackled problem.

Finally, Mastrolilli and Blum [192, 30] present a first theoretical work that shows the usefulness of incorporating bounding information for the purpose of constructing solutions in a parallel and non-independent way within metaheuristics.

Chapter 6
Hybridization Based on Complete Solution Archives

This chapter illustrates how a metaheuristic—in particular, evolutionary algorithms—can profit by cross-fertilization with basic principles of branch-and-bound: We extend the metaheuristic by a complete solution archive that stores all considered candidate solutions organized along the principles of a branch-and-bound tree. The approach is particularly appealing for problems with expensive evaluation functions or a metaheuristic applying indirect or incomplete solution representations in combination with non-trivial decoders. Besides just avoiding re-evaluations of already considered solutions, a fundamental feature of the solution archive discussed here is its ability to efficiently transform duplicates into typically similar but guaranteed new solution candidates. Thus, the solution archive can be seen to provide a kind of "informed mutation". From a theoretical point of view, the metaheuristic is turned into a complete enumerative method without revisits, which is in principle able to stop in limited time with a proven optimal solution. Furthermore, the approach can be extended by calculating bounds for partial solutions possibly allowing us to cut off larger parts of the search space. In this way the solution-archive-enhanced metaheuristic can also be interpreted as a branch-and-bound optimization process guided by principles of the metaheuristic search.

The chapter is structured as follows. Section 6.1 starts with the motivation and general idea of this kind of complete solution archives. After having a look at related work in the literature where solution archives have been used in conjunction with metaheuristics, Section 6.1.2 introduces the concept for problems with binary variables. Section 6.2 then studies this approach on the prominent Royal Road and NK landscape benchmark problems as a proof of concept. Practically more meaningful and more tailored versions of complete solution archives are then described in Section 6.3 for the Generalized Minimum Spanning Tree Problem, which we already considered in Chapters 2 and 4, when discussing decoder-based approaches and large neighborhood search, respectively. These solution archive variants will also be extended by bounding strategies, and in various experiments we will see the performance boost they provide.

6.1 General Idea

Most metaheuristics share a common weakness: Candidate solutions might in general be revisited multiple times, especially when a heuristic search has converged and emphasis is on exploitation rather than exploration. Such revisits and repeated evaluations of these solutions waste precious computing time. Typically this aspect is neglected as it is assumed that in a well-configured algorithm only a relatively small portion of all created candidate solution are duplicates. However, there are cases where such revisits may severely reduce a metaheuristic's effectivity: Especially when solution evaluations are expensive, possibly (1) because complex simulations need to be done or (2) when an indirect or incomplete solution representation is used in conjunction with a non-trivial decoder (see Chapter 2) and the search space is relatively small, it obviously makes sense to avoid or at least reduce repeated considerations of the same solutions.

The above is especially true for evolutionary algorithms (EAs, see Section 1.2.8) with a relatively small population size and variation operators that do not introduce much innovation, maybe because the emphasis lies in a later phase of the EA on a "fine-tuning". In the extreme case, the population's diversity drops strongly and the EA gets stuck by creating almost only duplicates of a small set of so far leading candidate solutions, called *super-individuals*. In such a situation of premature convergence, it becomes obvious that the heuristic search is not performing well anymore, and something must be changed in the EA's setup. Various kinds of population management strategies, such as accepting only solutions for a new generation that are sufficiently different from the others, have been suggested in the literature to counteract premature convergence, see, for example, [257, 193]. However, these mechanisms only consider duplicates w.r.t. the current population and *reject* them instead of more effectively *avoid their creation*.

We aim here at a solution archive that

- *stores all candidate solutions* considered throughout the whole metaheuristic run
- in a *memory-efficient* way,
- allows a *fast check whether or not a solution is already contained* (i.e., has already been evaluated),
- and provides an efficient function for *transforming a contained solution into a usually similar, guaranteed new solution ("informed mutation")*.

This archive is used within an EA or other metaheuristics to check and possibly transform each newly derived candidate solution before it is evaluated.

Note that such an archive-extension may in principle even turn the metaheuristic into a complete optimization approach that is guaranteed to find an optimal solution in limited time. This is trivially the case when the problem's search space is finite and the heuristic search has considered all possible solutions, which can also be efficiently recognized by the archive. This property, however, is usually more of theoretical interest, since in practice the heuristic search will typically be terminated much earlier.

6.1.1 Related Work

Looking at other metaheuristics, there is the popular *Tabu Search* (TS) [111] that makes explicit use of a memory, usually called *tabu list*, which keeps track of the search progress in order to avoid cycling; see Section 1.2.5. Different kinds of memories are used, but typically only attributes of recently performed moves are recorded and their reversal is forbidden for some time because of efficiency reasons. Thus, TS usually also does not entirely avoid revisits. Only a few TS approaches exist where entire solutions are directly or indirectly recorded in memory, and precisely those moves that would yield revisits are forbidden during the remaining search. The *reverse elimination method* [111] is one such example to realize what may be called *strict tabu search*. It is, however, relatively slow as at iteration n it requires a computation of order $O(n)$ to check whether or not a move is allowed. Therefore, Battiti and Tecchiolli [13] suggest to use classic *hashing methods* or a *digital tree* [160] instead, by which the essential insertion and find operations can be performed in (expected) time $O(l)$ with l being the size of the representation.

Battiti and Tecchiolli [13] further argue that such strict TS approaches might not work well in general as they often converge very slowly for problems where the local optima are surrounded by large *basins of attraction*[1]. This slow convergence is related to a slow "basin filling" effect. Well-tuned attribute-based approaches in which larger parts of the search space are temporarily disallowed are therefore typically more effective. In addition, global optima might even become unreachable because of the creation of barriers consisting of already visited solutions.

Focusing again on EAs, these negative aspects also require careful attention when extending an EA with a complete solution archive to avoid revisits. However, the possible implications are obviously less critical, as, in contrast to TS, an EA also includes other mechanisms such as recombination and mutation for diverting the search and jumping over barriers.

In the context of EAs, solution archives have been used more generally for a variety of different purposes. Occasionally, elite solutions are explicitly stored in an archive in order to re-introduce them later again, see, for example, [106]. In particular EAs for multiobjective optimization frequently use solution archives for storing non-dominated Pareto optimal solutions [75]. Obviously, the goals in such approaches are to avoid losing good solutions, which is somehow opposite to what we aim for in this chapter.

The idea of *caching* objective values of visited solutions to avoid (or reduce) re-evaluations is quite natural when the solution evaluation is expensive. Among others, Kratica [166] and Ronald [257] described such approaches using hash tables, while Louis and Li [180] suggest to store solutions in a binary search tree. Depending on the time effort of the objective function and the number of revisits,

[1] Remember, in this context, that the basin of attraction of a local optimum is defined as the set of all solutions for which a pre-defined local search method terminates in the local optimum when started from such a solution (see also Section 1.1.2).

computation times can be lowered substantially. However, revisits are not prevented or rejected in these methods.

Aiming at *completely avoiding revisits* in an EA for continuous optimization problems, Yuen and Chow [301] suggested to use an archive based on a k-d tree data structure for storing all visited solutions. When encountering a revisit, the corresponding solution is mutated in a special way in order to always derive a new, yet unvisited solution. This approach actually comes close to the concept we will consider in the following. Differences are, however, that we concentrate on a discrete search space and utilize a *trie* data structure resembling an explicitly stored branch-and-bound tree. This trie-based complete solution archive was originally proposed by Raidl and Hu in [245], from where we also adopt the experimental results.

6.1.2 Trie-Based Solution Archive for Binary Strings

First, we restrict ourselves to problems where solutions are encoded as binary strings of length l and feasibility of solutions is not an issue, i.e., the feasible search space corresponds to $\{0,1\}^l$. In this case, we use a *binary trie* as data structure for our archive. We will later expand the concept also to other representations.

Tries are data structures that are typically used to efficiently store a possibly very large set of strings [123] such as words in language dictionaries for spell checking. Different variants of tries exist, but they all have in common that the computational complexity of the insert and find operations essentially only depends linearly on the length of the respective word and not on the number of already stored words, i.e., these operations can be implemented in time $O(l)$. In comparison, balanced search trees would require $O(l \log n)$ time for these operations and typically also require significantly more memory. Hash tables are w.r.t. insert and find in the expected case asymptotically similarly efficient as binary tries, i.e., they also only require $O(l)$ time, but they do not allow an efficient realization of a *transform* operation that modifies an already stored solution into a similar new one. Note that *digital trees* [160] are also similar to binary tries.

6.1.2.1 Basic Binary Trie

The basic version of our trie for the considered binary search space is a binary tree T of maximum height $l-1$. Each node t_i at level $i = 1,\dots,l$, with level one corresponding to the root node, has identical structure: It consists of just two entries $t_i.next[0]$ and $t_i.next[1]$ which either are pointers referring to successor nodes at the next level or are set to the flags (special values) *completed* or *empty*.

Each node of the trie refers to a certain part of the search space: Let $x = (x_1,\dots,x_l) \in \{0,1\}^l$ be a solution vector. The root node t_1 corresponds to the whole space $\{0,1\}^l$, and for each node t_i at level $i = 1,\dots.l$ the following holds: Entries $t_i.next[0]$ and $t_i.next[1]$ of trie node t_i partition the corresponding search space into

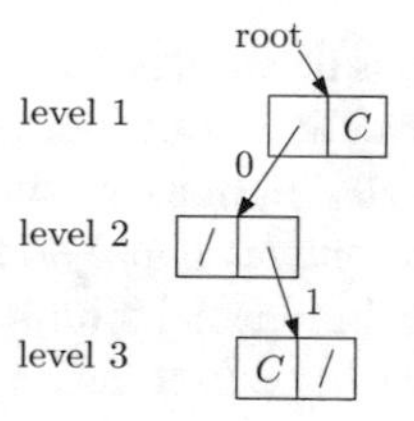

Fig. 6.1 A trie containing solutions 010, and all solutions with $x_1 = 1$, i.e., 100, 101, 110, and 111 (from [245])

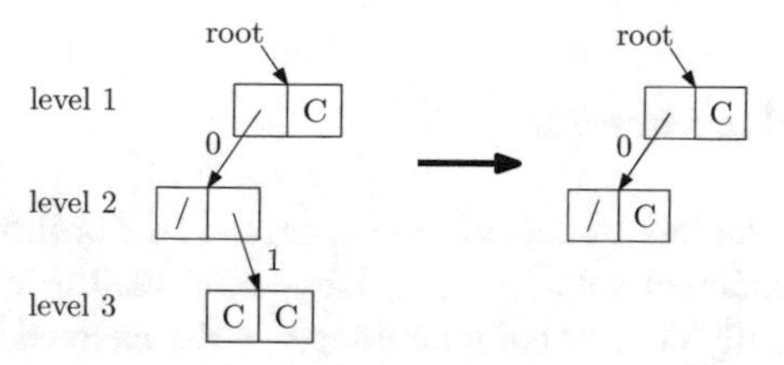

Fig. 6.2 Pruning the subtrie containing solutions 010 and 011 (from [245])

the two subspaces containing those vectors with $x_i = 0$ and $x_i = 1$, respectively. Thus, the i-th variable of a vector x decides whether to go "left" or "right" at level i.

An entry of *empty* indicates that none of the solutions lying in the corresponding subspace is yet contained in the trie, while an entry *completed* refers to the case that all corresponding solutions are contained, i.e., the corresponding subspace has been completely searched by the metaheuristic. Figure 6.1 shows an exemplary trie holding the solution 010 and all solutions with $x_1 = 1$; 'C' denotes the *completed*-flag and '/' the *empty*-flag.

Note the close correspondence of this trie data structure to a branch-and-bound tree: For the binary search space, it is most natural to realize a branch-and-bound by always branching over one variable, i.e., for each subproblem selecting a not yet fixed variable x_i and fixing it to zero and one to obtain two disjoint subproblems (subspaces), respectively. A major difference is, however, that in branch-and-bound this tree is usually not explicitly stored.

To check whether or not a given solution x is stored in T, the lookup algorithm starts at the root and steps down the trie, always following $t_i.next[x_i]$ at each level i. If a *completed*-flag is reached during this process, x is contained in T; if *empty* is encountered, x has not yet been inserted. This lookup can thus be efficiently performed in time $O(l)$.

Adding new, not yet contained solutions works similarly, but new trie nodes must possibly be created when encountering an *empty* flag, and after considering the last variable x_l a corresponding *completed*-flag is stored.

An important principle to let the trie not become unnecessarily large is to prune a subtrie when all solutions corresponding to it have been added, i.e., if $t_i.next[0] = t_i.next[1] = completed$, the respective pointer to t_i in the previous level is replaced by a *completed*-flag and the subtrie is deleted; see Fig. 6.2, where 011 has additionally been inserted and the node at level three can successively be pruned.

Note that in the case when all solutions of the whole search space have been added, the root pointer becomes *completed* and no trie nodes remain. Thus, this data structure is also able to indicate this special situation most efficiently in time $O(1)$. In the metaheuristic, this check can be used as an additional termination criterion, stopping the search when the search space has been completely exhausted and thus a guaranteed optimal solution has been found (provided that the metaheuristic keeps track of the so far best considered solution).

6.1.2.2 Transforming Solutions

The most important feature of the solution archive is the ability to efficiently transform an already contained solution x, which would lead to a revisit in the metaheuristic, into a usually similar but guaranteed so far unconsidered solution x'. By "similar" we mean in this context that the Hamming distance between x and x' is low. Considering the example of Fig. 6.1, if the metaheuristic's variation operator yields $x = 010$ again, this solution is transformed into $x' = 011$, which is then inserted in T leading to the situation shown in Fig. 6.2.

More generally, the basic concept of the transform operation is to go back to some previous node at the search path of x whose alternate branch $p.next[1 - x_i] \neq$ *completed*, i.e., contains at least one yet unconsidered solution. Here, at this *deviation position*, we set $x_i = 1 - x_i$ and go down this alternate subtrie following the remaining variables of x whenever possible, i.e., unless a *completed*-flag is encountered. In the latter case, we choose again the alternate branch that must contain at least one unconsidered solution as otherwise the node would have been pruned already.

Concerning the selection of the deviation position, we consider two transformation variants:

Deepest Position Transformation (DPT): In this basic variant the last, i.e., deepest possible, deviation position is always chosen. While this method is most straightforward as it corresponds to a classic backtracking in a depth-first search, it has the disadvantage of yielding a strong bias towards modifying variables with higher indices while keeping variables with lower indices unchanged. In fact, only when large portions of the search space have already been covered will variables with lower indices be considered.

Random Position Transformation (RPT): To substantially reduce the bias of DPT, a random choice from all feasible deviation positions is made in the RPT variant. Otherwise, this method follows the same principle as DPT.

Both transformation variants are stated in pseudo-code in Algorithm 27. It first searches for x, then goes back to the deviation point and down again in order to insert the transformed solution. As a feasible deviation point can always be selected efficiently, no further backtracking is necessary, and thus, this algorithm can be efficiently implemented in time $O(l)$.

Algorithm 27 Transform Solution

```
given: solution x
p = root
devpoints = ()
{search for x storing possible deviation positions in devpoints}
i = 1
while i ≤ l ∧ p ≠ completed do
    if p.next[1 − x_i] ≠ completed then
        devpoints = devpoints ∪ (i, p)
    end if
    p = p.next[x_i]
    i = i + 1
end while
if variant = DPT then
    (i, p) = last entry of devpoints {go back to last feasible deviation position}
else
    (i, p) = random entry from devpoints {go back to a random deviation position}
end if
x_i = 1 − x_i {actually deviate by flipping variable x_i}
while i ≤ l do
    if p.next[x_i] == completed then
        x_i = 1 − x_i
    end if
    if p.next[x_i] == empty then
        p.next[x_i]=new trie node (empty, empty)
    end if
    q = p
    p = p.next[x_i]
    i = i + 1
end while
q.next[x_i] = completed {insert transformed x in T}
prune trie nodes containing (completed, completed) bottom up
return x' = x
```

6.1.2.3 Randomizing the Trie

In the above introduced basic trie, nodes on level i are always associated with variable x_i, $i = 1, \dots, l$. This fixed association may have a significant weakness: Especially when using DPT for handling duplicates, we already observed that there is a strong bias towards variables with higher indices being changed more frequently than those with lower index. This bias typically results in repeated, rigid patterns of visited solutions in the search space. In general, when an intensive search around an incumbent solution has been performed and the trie is already moderately filled, the transform operation might need to flip not just the single variable at the devia-

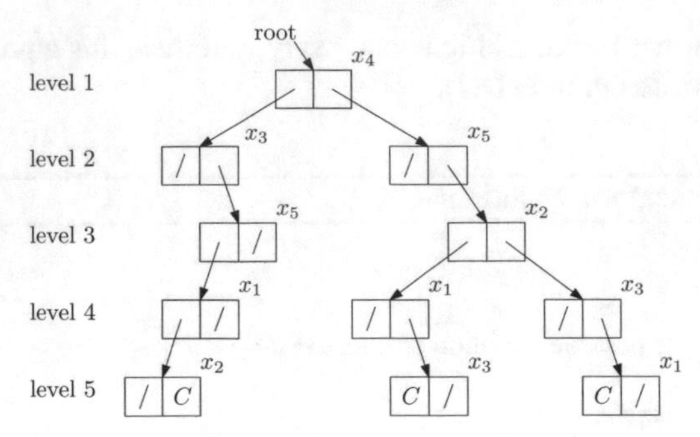

Fig. 6.3 A randomized trie containing 01100, 10011, 01111. Nodes are labeled by their associated variables (from [245])

tion point but more variables in the course of finding a yet unconsidered solution, resulting in larger Hamming distances. As DPT considers the positions always in strictly the same order, this effect is significantly amplified, and transformed solutions with larger distances, i.e., less similar solutions, are created earlier and/or more frequently.

As mentioned above, the RPT variant reduces this weakness by choosing the deviation position randomly, but it is still not able to entirely avoid a bias. As an alternative or additional improvement, we consider the randomization of the trie structure in the following way. The idea is to use individual, in general different, variable orders on different search paths of the trie. Trie nodes at depth i are no longer always related to variable x_i, but a deterministic pseudo-random (or hash) function is used to calculate the specific variable index j related to a given trie node t_i in dependence of the whole path from the root to t_i. Figure 6.3 illustrates this randomized trie variant.

Using a deterministic pseudo-random function has the advantage that no additional information needs to be stored in the trie, but the relevant variable related to a given trie node can always be efficiently determined. It is important, however, to carefully choose this pseudo-random function. For example, classic Park-Miller random number generators are unsuitable as there are correlations between input and output data. In the experiments below we used the "pseudo data-encryption standard" algorithm *ran4* [236].

6.2 Application to Royal Road Functions and NK Landscapes

To get an impression of the impact of the above presented trie-based solution archive, we first follow [245] and apply the concept within a simple genetic algorithm (GA) for solving two classic benchmark problems in the binary space: Royal Road functions and NK landscapes.

Table 6.1 Considered GA variants

GA variant	solution archive	transform operation
STD	–	– (classic mutation)
TBD	basic trie	deepest position (DPT)
TBR	basic trie	random position (RPT)
TRD	randomized trie	deepest position (DPT)
TRR	randomized trie	random position (RPT)

The GA is steady-steady, i.e., it creates at each iteration only one new solution which always replaces the population's worst solution. For deriving a new solution, tournament selection with replacement is applied twice to select two parent solutions. These are recombined by a classic single point crossover. Mutation is performed on the new solution by flipping each variable with a probability of $1/l$; i.e., in the expected case one variable is flipped.

When the solution archive is attached, classic mutation is turned off and replaced by the transform operation of the archive when already visited solutions are obtained from recombination. Initial solutions are created uniformly at random. In the experiments below, the population size was 100 and the tournament selection group size 10. We compare the GA variants and trie configurations summarized in Table 6.1.

To make the comparison to the standard GA without solution archive (i.e., the STD-variant) fairer from a practical perspective, a classic duplicate elimination strategy, as described, for example, in [257], is applied in this case, i.e., a newly derived solution is only accepted in the population if it is different from all other solutions therein and discarded otherwise. Thus, there are also never any duplicates in the current population. The necessary duplicate-checks are efficiently performed by means of a hash table for the population. Tests without any duplicate elimination and without the suggested archive yielded consistently clearly worse results, and we therefore do not include them here. All experiments were performed on a Pentium 4, 2.8 GHz PC.

6.2.1 Results for Royal Road Functions

Royal Road functions are a frequently used class of artificially defined benchmark problems. Mitchell et al. [198] originally proposed them to evaluate GAs w.r.t. their recombination operators. These functions are defined for binary strings $x \in \{0,1\}^l$ on a set of hierarchically created *schemas* $S = \{s_1, s_2, \ldots, s_n\}$. A schema is a subspace of the whole search space where some variables are fixed to certain values and others remain open. For example, the schema *10** refers to all solutions with $x_2 = 1$, $x_3 = 0$ and x_1, x_4 and x_5 being either zero or one. The *order* $O(s)$ of a schema s is its number of fixed variables. The objective of the Royal Road functions is to maximize $\sum_{s \in S} o(s)\sigma_s(x)$, where $\sigma_s(x)$ is the binary indicator-function yielding one if x matches s and zero otherwise.

Table 6.2 Royal Road functions: Average results from solution-archive-enhanced GA variants for different b and r settings (from [245])

instance		STD		TBD		TBR		TRD		TRR	
b	r	$\overline{obj}$	σ	$\overline{obj}$	σ	$\overline{obj}$	σ	$\overline{obj}$	σ	$\overline{obj}$	σ
3	4	36.00	0.00	36.00	0.00	36.00	0.00	36.00	0.00	36.00	0.00
4	4	48.00	0.00	48.00	0.00	48.00	0.00	48.00	0.00	48.00	0.00
5	4	60.00	0.00	60.00	0.00	60.00	0.00	59.30	4.95	60.00	0.00
6	4	62.88	18.71	64.32	16.58	68.52	11.94	68.28	12.94	68.64	11.51
3	8	96.00	0.00	96.00	0.00	96.00	0.00	96.00	0.00	96.00	0.00
4	8	124.40	14.39	126.80	8.49	124.40	14.39	125.60	11.88	128.00	0.00
5	8	103.50	45.41	115.40	43.22	115.20	42.89	110.60	42.64	125.30	44.01
6	8	73.80	35.16	92.64	36.95	81.72	51.38	77.76	38.06	81.84	44.88
3	16	206.28	45.43	215.76	41.31	217.68	40.12	226.98	32.60	219.54	38.92
4	16	148.08	43.99	160.08	55.48	166.16	63.70	168.00	66.07	153.44	55.26
5	16	106.50	37.95	100.00	38.69	104.50	38.61	93.00	32.09	102.30	41.46
6	16	70.44	30.52	79.44	41.81	74.16	29.34	74.52	29.50	82.68	35.01

The set of schemas S is created by starting with a basic building block of a given order b, which is repeated r times at subsequent positions to obtain r initial schemas. These are then pairwise combined forming larger and larger building blocks and corresponding schemas in a hierarchical fashion until a final building block covering all positions is obtained. The parameters b and r are called *base* and *multiplier*, respectively, and in our case $l = b \cdot r$. For more details on this schema creation we refer to [198].

Table 6.2 lists results on differently parameterized Royal Road functions. All runs are terminated after 1000 iterations. Average final objective values ($\overline{obj}$) and corresponding standard deviations (σ) of 100 independent runs are printed for each test case. Best values are highlighted. Table 6.3 further shows the following total results over all considered Royal Road functions: Average objective values, average elapsed CPU-times when the best solutions were identified, the number of solution transformations (i.e., avoided revisits), and for each pair of GA variants the error probability (p-value) of a one-sided Wilcoxon rank sum test for the hypothesis that the GA variant corresponding to the row performs better than that of the column. Cases where the error probability is less than 5% are highlighted.

We can observe that the GA without the trie archive performs significantly worse in general. Among the different trie-variants, performance differences are rather small, but using the random deviation transformation and the randomized trie structure leads to noticeable improvements.

6.2.2 Results for NK Landscapes

NK landscapes are another popular benchmark suit for binary representations. They were introduced by Kauffman [153]. The objective is to maximize the function

Table 6.3 Royal Road functions: Averages over all instances, CPU-times, and p-values of Wilcoxon rank sum tests for each pair of GA variants (from [245])

alg	$\overline{obj}$	time	transf.	p(STD)	p(TBD)	p(TBR)	p(TRD)	p(TRR)
STD	92.96	0.02s	–	–	0.9953	0.9942	0.9677	0.9996
TBD	99.54	0.03s	312	0.0047	–	0.3854	0.2901	0.6849
TBR	99.36	0.05s	297	0.0058	0.6150	–	0.3609	0.7099
TRD	98.67	0.03s	307	0.0323	0.7102	0.6394	–	0.9122
TRR	100.15	0.05s	301	0.0004	0.3154	0.2904	0.0879	–

Table 6.4 NK landscapes: Average results of the solution-archive-enhanced GA variants for different settings of K over $N \in \{20, 50, 100, 300\}$ (from [245])

	STD		TBD		TBR		TRD		TRR	
K	$\overline{obj}$	σ	$\overline{obj}$	σ	$\overline{obj}$	σ	$\overline{obj}$	σ	$\overline{obj}$	σ
1	0.7090	0.0285	0.7089	0.0288	0.7086	0.0288	0.7092	0.0286	0.7089	0.0288
2	0.7351	0.0220	0.7354	0.0217	0.7364	0.0219	0.7361	0.0217	0.7365	0.0216
5	0.7603	0.0222	0.7623	0.0222	0.7611	0.0219	0.7609	0.0219	0.7633	0.0211
6	0.7628	0.0232	0.7631	0.0239	0.7641	0.0227	0.7645	0.0226	0.7649	0.0225
7	0.7595	0.0223	0.7583	0.0217	0.7592	0.0211	0.7607	0.0216	0.7600	0.0219
8	0.7567	0.0241	0.7583	0.0249	0.7582	0.0244	0.7597	0.0232	0.7608	0.0245
9	0.7545	0.0219	0.7526	0.0239	0.7557	0.0212	0.7560	0.0244	0.7571	0.0228
10	0.7505	0.0231	0.7522	0.0247	0.7507	0.0269	0.7543	0.0248	0.7528	0.0267

Table 6.5 NK landscapes: Average objective values over all instances and p-values of Wilcoxon rank sum tests for each pair of GA variants (from [245])

alg	$\overline{obj}$	σ	time	transf.	p(STD)	p(TBD)	p(TBR)	p(TRD)	p(TRR)
STD	0.7485	0.0290	3.47s	–	–	0.6851	0.9420	1.0000	1.0000
TBD	0.7489	0.0295	3.53s	52215	0.3149	–	0.8358	0.9999	1.0000
TBR	0.7492	0.0294	3.69s	49602	0.0580	0.1643	–	0.9994	1.0000
TRD	0.7502	0.0294	3.48s	54003	0.0000	0.0001	0.0006	–	0.8472
TRR	0.7506	0.0298	3.79s	50718	0.0000	0.0000	0.0000	0.1529	–

$$F(x) = \frac{1}{N} \sum_{i=1}^{N} f_i(x_{i_1}, x_{i_2}, \ldots, x_{i_K}) \tag{6.1}$$

with $x \in \{0,1\}^N$. Each subfunction f_i takes variable x_i and K neighboring variables $x_{i_1}, \ldots, x_{i_K}$ into account and returns a value in $[0,1]$ according to a random value table. Hence there are N tables of size 2^{K+1} from which the values are taken. With increasing K, the coupling between particular variables rises and the problem becomes more complex.

Two variants exist for choosing the neighboring genes $x_{i_1}, \ldots, x_{i_K}$: the *adjacent neighbors* method, where these genes are the closest ones to x_i, and the *random neighbors* method, where they are selected randomly from all $x_1, \ldots, x_N$. We consider here the NP-hard latter case.

Parameter N was set to 20, 50, 100 and 300 and K to 1, 2, 5, 6, 7, 8, 9 and 10. For each combination, we performed 50 independent runs and each run was terminated after 10 seconds. Thus, the time overhead introduced by the archive is taken into account here. Since the final objective values for different N but the same K are relatively similar, we give here accumulated results grouped by K. Table 6.4 shows these average final solution values and corresponding standard deviations for the standard GA and the four trie-enhanced variants. Though the average objective values here lie close together, the advantage of the trie becomes obvious. In particular, the GA variant using the random deviation transformation together with the randomized trie structure was able to generate best average results most of the time. This becomes even more evident in Table 6.5 where average values over all values of K together with pairwise error probabilities of Wilcoxon rank sum tests are presented analogously to Table 6.3.

6.3 Application to the Generalized Minimum Spanning Tree Problem

The above exemplary application of the solution archive in a GA for the Royal Road functions and NK landscapes serves as a proof of concept. The obtained gains are statistically significant in the studied case of a simple GA, but might be questionable when considering that the GA may also be enhanced by other techniques, such as other population management techniques or some embedded local search.

As another, more practical example, where the impact of a solution archive will turn out to be more substantial, we consider again the *Generalized Minimum Spanning Tree* (GMST) problem introduced in Section 2.2. Remember that we used this problem to illustrate decoder-based variable neighborhood search approaches as well as a large neighborhood search in Section 4.3. In contrast to Sections 2.2 and 4.3 we consider now an evolutionary algorithm framework, as an EA integrates more naturally with the solution archive concept. The following section presents two different trie variants tailored to the GMST's *Node Set Based* (NSB) and *Global Edge Set Based* (GESB) representations, respectively. Section 6.3.3 describes the EA framework, Sections 6.3.4 and 6.3.5 introduce bounding extensions to the solution archives by which we will be able to prune the search space in a way directly related to branch-and-bound, and finally, experimental results are presented in Section 6.3.6. These sections follow the works from Hu and Raidl [140, 141].

6.3.1 Solution Archive for the Node Set Based Representation

In the NSB representation, a candidate solution is encoded by a vector of nodes $p = (p_1, \ldots, p_r) \in V_1 \times \ldots \times V_r$ to be connected and the decoder determines via a classic *Minimum Spanning Tree* (MST) algorithm the edges.

Obviously, we cannot directly use the binary trie from Section 6.1.2.1 but have to adapt it appropriately. Considering a vector of nodes p, it appears natural to fix at

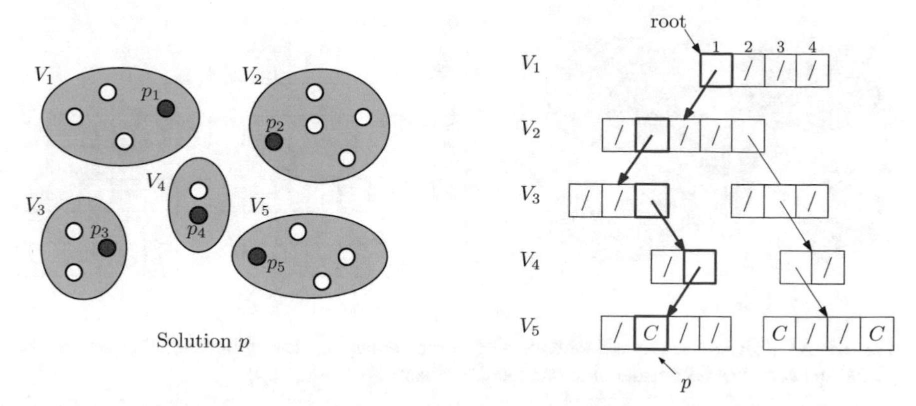

Fig. 6.4 An NSB solution p is stored in an NSB trie containing already two other solutions. The emphasized arcs indicate the search path (from [141])

each level of the trie the node selected for one cluster. Then, our trie has a maximal height of $r-1$, as r clusters must be considered. Each trie-node at level $i = 1,\ldots,r$ is associated with cluster V_i and contains entries $next[j]$, $j = 1,\ldots,|V_i|$. Thus, we do not have a binary trie anymore but each node at level $i < r$ can have up to $|V_i|$ successors. Each entry $next[j]$ is either a reference to a trie-node on the next level, a *completed*-flag, or an *empty*-flag. As in the binary trie, *empty* means that none of the solutions in the subspace corresponding to this subtrie is contained and thus has been considered by the EA, while the entry *completed* indicates that the whole subspace has already been completely explored.

The insertion procedure starts at the root and follows at each node at level i the entry that corresponds to the value of p_i, i.e., the node selected for cluster i. In the trie-node of the last level, $next[p_r]$ is set to *completed*, indicating that this solution is now contained. Figure 6.4 shows an example of how a solution p is stored. Since we want to keep the trie as compact as possible, subtries where all solutions have been visited are pruned again. This is done by removing trie-nodes that only contain *completed*-flags and changing the respective entry in the predecessor node (or ultimately the pointer to the root node) into *completed*.

The transformation function for producing a new solution from an already visited one essentially works as described for the binary trie in Section 6.1.2.2: A deviation point is chosen from all nodes at the search path having at least one other entry that is not marked as *completed*. One of the node's uncompleted entries is randomly selected, and the procedure follows this subtrie down considering the remaining variables from p whenever possible, i.e., unless a *completed* is encountered again. In the latter case an alternative not-yet-completed branch is chosen randomly. As with the binary trie, this procedure is guaranteed to efficiently find a new solution by going down from the deviation point only once. Concerning the selection of the deviation point, we follow here always the random position transformation strategy due to its advantage of being substantially less biased than the deepest position strat-

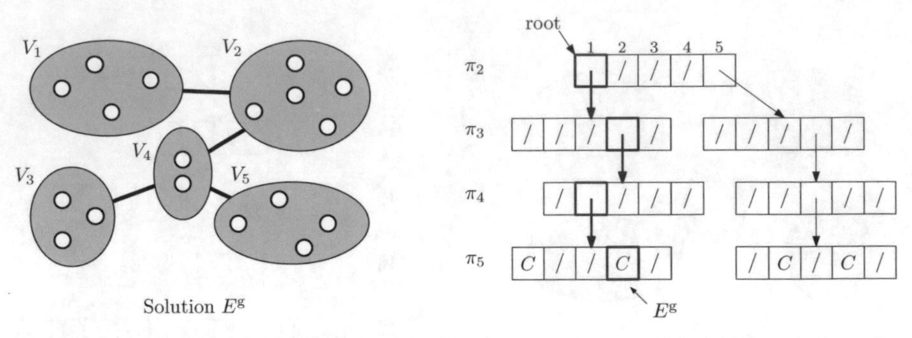

Fig. 6.5 A GESB solution E^g is stored in a GESB trie containing already three other solutions; the global spanning tree is considered to be rooted at cluster V_1 (from [141])

egy. While a simple search runs in time $O(r)$, insertion and transformation require time $O(r \cdot d_{\max})$ with $d_{\max}$ being the maximum number of nodes in a cluster due to the larger nodes.

6.3.2 Solution Archive for the Global Edge Set Based Representation

For the GESB representation an effective data structure for the solution archive is less obvious, as a feasible solution must correspond to a spanning tree. We consider the global spanning tree as rooted in cluster V_1 and direct all global edges in such a way that we obtain an outgoing arborescence, i.e., there is a directed path from V_1 to each other cluster. Then, each cluster except V_1 has a unique predecessor, and these predecessors represent the global spanning tree in a unique way. The trie T_{GESB} dedicated to the GESB encoding is thus based on the *predecessors vector* $\pi = \{\pi_2, \ldots, \pi_r\}$, where $\pi_i, i = 2, \ldots, r$ is the index of the predecessor of cluster i. The trie has maximal height $r-2$, and each trie-node at level $i = 1, \ldots, r-1$ corresponds to the predecessor π_{i+1} and contains entries $next[j]$, $j = 1, \ldots, r$. Figure 6.5 shows an example of how a global spanning tree is stored in T_{GESB}.

Inserting, searching and transforming a solution in this trie follows the same principles as before. While searching can be implemented in time $O(r)$, inserting requires time $O(r^2)$ since the nodes have now size $O(r)$.

Regarding solution transformation, a special aspect needs to be considered: As the global edge set must form a spanning tree, only a subset of all possible assignments $\pi \in \{1, \ldots, r\}^{r-1}$ is feasible. Changing a variable π_i to some arbitrary other value in $\{1, \ldots, r\}$ may result in an invalid disconnected graph containing a cycle. We therefore check each transformed solution and if it is infeasible, we apply a simple depth-first-search for the next feasible solution as a repair strategy. In this search, all encountered infeasible solutions are marked in the trie as *completed*.

We remark here that instead of basing the trie on predecessor vectors, one might use Prüfer sequences [239]. A Prüfer sequence is a vector in $\{1,\ldots,r\}^{r-2}$ which always represents a unique spanning tree on r nodes, i.e., Prüfer sequences provide a bijection to all spanning trees. In this case, no repairing is necessary and each transformation runs in time $O(r^2)$. A major disadvantage of this approach, however, is that a small change in a Prüfer sequence frequently yields a very different spanning tree [246], and thus, the transformation would in general not produce similar solutions anymore but correspond more or less to a "random jump" in the search space. Therefore we stick to the predecessor-based variant, despite its necessity to deal with infeasible solutions.

6.3.3 The Archive-Enhanced Evolutionary Algorithm for GMST

As in our experiments concerning the Royal Road functions and NK landscapes, we use a classic steady-state EA where the archive is consulted each time after a new solution is generated by recombination and mutation. The major difference from an architectural point of view is only the fact that we make use of the two incomplete representations NSB and GESB and the corresponding decoders. For each (complete) solution in the population we store its vector of selected nodes p together with the global edge set E^g, with the latter further encoded as predecessor vector π.

To select parental solutions for the variation operations we apply binary tournament selection. Each new candidate solution obtained by recombination and mutation and possibly transformed by one or both of the solution archives in order to avoid duplicates always replaces the worst solution in the population.

6.3.3.1 Variation Operators

Recombination is performed by first deciding randomly which representation to use, i.e., either NSB or GESB.

In the case of NSB, the node vectors of the two parental solutions are recombined by a classic uniform crossover, and optimal edges are determined by the MST-algorithm based decoder.

In the case of GESB, the spanning tree aware *edge recombination* from [246] is applied to the parents' global edge sets E^g. This recombination first adopts all edges both parents have in common and then augments this set by choosing additional edges from either parent following a randomized Kruskal MST algorithm until a feasible spanning tree is obtained. Finally, the solution is completed again by selecting optimal nodes via the DP-based decoder from Section 2.2.3.

Mutation is based on the same considerations, i.e., it is decided randomly which representation to use. In the case of NSB, one node in one cluster is randomly exchanged, while in GESB, a random edge exchange takes place, ensuring, however, that the obtained global edge set represents a spanning tree again. In both recombination variants, the respective decoders are again applied to optimally augment partial solutions.

After applying recombination and mutation, the obtained candidate solution is checked and possibly transformed by both solution archives, first the NSB archive and then the GESB archive. Of course, the respective decoder is also applied in case of a transformation in order to obtain a complete solution again. Since a transformation by the GESB archive will in general also yield a different node selection due to the reapplied decoder, the NSB and GESB archive checks and transformations are iteratively performed until no further transformation is necessary, i.e., a guaranteed not yet considered solution has indeed been found.

Note that mutation plays in this archive-enhanced EA only a minor role since duplicate solutions are transformed in the solution archive anyway. However, experiments indicated that it is still beneficial to include it in the described way with some low probability so that fewer duplicates originally arise and transformations are not required so often.

6.3.4 Pruning the NSB Trie by Bounding

In the following we extend the solution archive concept in a substantial way, adopting a major functionality of branch-and-bound: *bounding*. The correspondence of our trie data structures to an explicitly stored branch-and-bound tree has already been pointed out. It is therefore only natural to consider possibilities for pruning open subtries that cannot contain better solutions than the best solution found so far, instead of enumerating all contained solutions. We achieve this as in branch-and-bound by calculating bounds on the objective values that might be reached in the best-case for the partial solutions corresponding to intermediate trie nodes. If such a bound is worse than the objective value of the so far best found solution, the respective subtrie can be pruned, i.e., is set to *completed*. Note that as in branch-and-bound, this pruning is safe in the sense that it does not avoid finding in principle a proven optimal solution in the end. We now investigate such bounding extensions in detail for the NSB trie. Section 6.3.5 describes an corresponding approach for the GESB trie.

In order to possibly prune a certain uncompleted subtrie, a lower bound needs to be efficiently determined for the objective values of all solutions in the subspace of the search space corresponding to the subtrie. To obtain such a bound for a particular entry in a trie-node corresponding to cluster V_k, $k = 1, \ldots, r$, we consider an auxiliary graph $G^{\mathrm{NSB}} = (V^{\mathrm{NSB}}, E^{\mathrm{NSB}})$ which is defined as follows.

V^{NSB} is composed of two sets of clusters: the "fixed" clusters V^{f} for which a specific node has already been selected, i.e., p_i is known, and the set of "open" clusters V^{o}, i.e., those for which p_i is yet undefined; $i = 1, \ldots, r$.

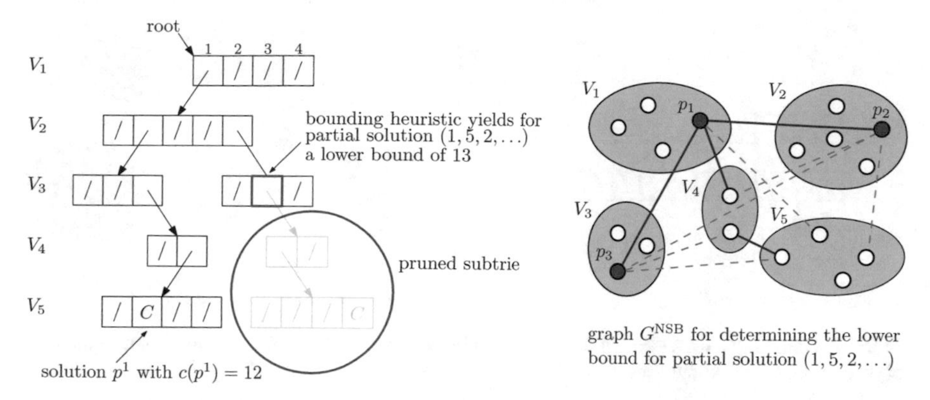

Fig. 6.6 Obtaining a lower bound for an NSB subtrie and pruning it as the bound is larger than the objective value of a known solution p^1 (from [141])

For the open clusters in V^{o} we relax the GMST condition that they must be connected using a single node, i.e., we allow arbitrary edges to any, even multiple, nodes in these clusters.

The edge set E^{NSB} is composed of three subsets: $E^{\text{NSB}} = E^{\text{ff}} \cup E^{\text{fo}} \cup E^{\text{oo}}$. Set E^{ff} contains connections among clusters of set V^{f}, i.e., $E^{\text{ff}} = \{(V_i, V_j) \mid V_i, V_j \in V^{\text{f}}\}$. The cost of such a connection corresponds to the actual edge cost $c(p_i, p_j)$. Set E^{fo} contains connections between a cluster of set V^{f} and a cluster of set V^{o}, i.e., $E^{\text{fo}} = \{(V_i, V_j) \mid V_i \in V^{\text{f}},\ V_j \in V^{\text{o}}\}$. The cost of each of those connections is $\min\{c(p(V_i), v) \mid v \in V_j\}$. Finally, E^{oo} contains connections among clusters of set V^{o}, i.e., $E^{\text{oo}} = \{(V_i, V_j) \mid V_i, V_j \in V^{\text{o}}\}$, and the corresponding costs are $\min\{c(u, v) \mid u \in V_i,\ v \in V_j\}$. A lower bound can now efficiently be obtained by applying a standard MST algorithm to G^{NSB}.

Figure 6.6 shows an example for this approach. Assume that a previously obtained solution $p^1 = (1, 2, 3, 2, 2)$ has objective value $c(p^1) = 12$. A lower bound for all solutions starting with $(1, 5, 2, \ldots)$ is calculated on the graph G^{NSB}. All shown lines represent the whole set of considered connections in E^{NSB} and the bold lines are those that are actually selected by the MST algorithm. Note that in the corresponding subgraph of the original graph G, neither does only one node have to be connected per cluster, nor is the subgraph necessarily connected. Assuming this bounding heuristic obtains a lower bound of 13 in our example, then there is no need to consider any further solutions in the corresponding subtrie, and consequently it is pruned.

Building up E^{NSB} can be computationally expensive if it is done every time from scratch for calculating a lower bound. However, the sets E^{fo} and E^{oo} for all possible solutions can be computed once in advance in $O(|V|^2)$ time during preprocessing. Therefore the time-complexity of the bound calculation for a partial solution is dominated by applying the MST algorithm on E^{NSB} requiring $O(r^2 \log r)$ time.

Nevertheless it would be a too large overhead in practice to do the bound calculation at each level when inserting a new solution or transforming a duplicate. On

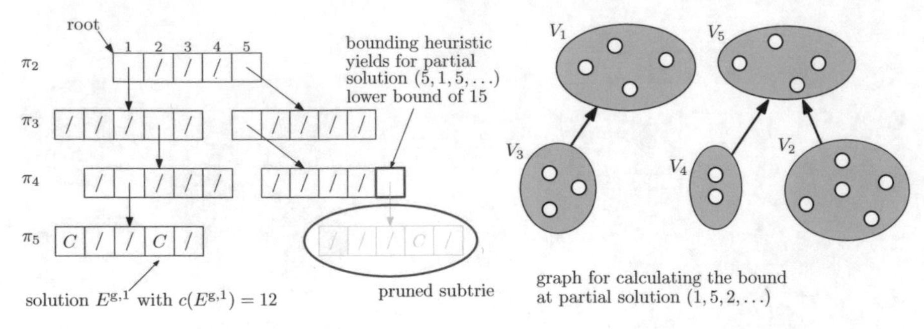

Fig. 6.7 Obtaining a lower bound for a GESB subtrie and pruning it as the bound is larger than the objective value of a known solution $E^{g,1}$ (from [141])

the one hand, being able to prune a subtrie at a high level (i.e., low depth) means that potentially more solutions are excluded at once; thus more space and time may be saved. On the other hand, the lower bounds for nodes at higher levels of the trie where relatively few clusters are fixed are less frequently tight enough to let this happen. In the preliminary experiments from [140] there was not a single case where such a pruning could be performed in the upper half of the trie. Therefore the bounding procedure is applied in the experiments during insertion and transformation on all accessed nodes of the lower half of the trie with a certain probability. Corresponding results will be shown in Section 6.3.6.

6.3.5 *Pruning the GESB Trie by Bounding*

The lower bound calculation for a subtrie in the GESB trie works as follows. The partially fixed predecessor vector $\pi = \{\pi_2, \ldots, \pi_r\}$ corresponding to the considered subtrie represents a global structure being in general a forest $F^{\mathrm{g}} = \bigcup_{i=1,\ldots,m} K_i^{\mathrm{g}}$ where $K_i^{\mathrm{g}} = (V_i^{\mathrm{g}}, T_i^{\mathrm{g}})$, $i = 1, \ldots, m$, are m connected global tree components of F^{g} which are not connected among each other; V_i^{g} denotes the set of clusters that are contained in component K_i^{g}. For every K_i^{g}, $i = 1, \ldots, m$, that is not a single cluster we apply the DP procedure also used as overall decoder for the GESB representation, see Section 2.2.3. As a result we obtain for each K_i^{g} the selection of nodes to be spanned for the clusters in V_i^{g} minimizing the connection costs. After all global tree components K_i^{g} are independently processed in this way, they are connected by using the same MST heuristic as in the case of the NSB representation. For this purpose the cheapest connections between all components are considered and the condition to connect exactly one node in each cluster is relaxed again. Since each component K_i^{g} is connected in the cheapest way, we obtain a lower bound.

Figure 6.7 shows an example for this procedure. Assume we want to compute the lower bound for the solution space starting with the partial predecessor vector $(5, 1, 5)$. Forest F^{g} contains two global tree components: K_1^{g} with clusters V_1 and V_3

and K_2^g with clusters V_2, V_4, and V_5. Both components are independently processed by the DP-based decoder yielding the cheapest connections and selected nodes. Assuming we obtain a lower bound of 15 and a previously known solution $E^{g,1}$ having objective value 12 exists, the corresponding subtrie can be pruned.

The complexity of determining lower bounds in the GESB trie is higher than for doing so in the NSB trie. The cheapest connections between clusters and/or nodes are the same as those already precomputed for the NSB representation. However, an additional effort is necessary for decoding the m global tree components. In total this requires up to $O(|V|^2)$ time (it will not take longer than decoding a complete GESB solution).

An alternative would be to not decode the components exactly but to use the cheapest connections between clusters as lower bounds instead. Such an approach is much faster, but the obtained bounds are usually very weak, and pruning can take place only rarely, as experiments in [141] indicated.

GESB's lower bounds are frequently tighter than those from the NSB representation. On the one hand this is due to the more sophisticated bounding calculation where runtime is not spent in vain. On the other hand decisions in the GESB representation are more powerful per se, since fixing a connection between two widely separated clusters typically has a more direct and larger impact on solution quality than fixing a less suitable node in a cluster in the NSB representation. While pruning in the NSB representation is typically only possible in the lower half of the trie, this does not hold for the lower bounds calculated for the GESB trie. Experiments indicated that pruning can happen on essentially any level with a reasonably high probability. As in the NSB trie, the bounding procedure is applied for any encountered node during insertion and transformation with a certain probability.

6.3.6 Results

In the following experiments the TSPlib[2] instances with geographical center clustering according to [95] are considered. For each instance and each algorithm variant, 30 independent runs were performed in order to obtain final average objective values and corresponding standard deviations. New candidate solutions are derived by binary tournament selection, recombination and performing mutation with a probability of 10%. The population always consisted of 100 solutions, and a new solution always replaces the worst solution in the population. All experiments reported here were performed on a single core of an Intel Core 2 quad CPU with 2.4 GHz.

We consider different variants of the EA, applying either only the NSB archive, the GESB archive, or both together; results are adopted from Hu and Raidl [141]. The bounding calculation to prune subtries is either applied with a probability of 5% at each accessed trie node and in the case of the NSB archive only at the lower half at levels $\lceil r/2 \rceil, \ldots, r$, or turned off, as will be indicated.

---

[2] http://elib.zib.de/pub/Packages/mp-testdata/tsp/tsplib/tsplib.html

Table 6.6 Results of different EA variants for the GMST problem on TSPlib-based instances

		no archive		NSB archive		GESB archive		both archives		both+bounding	
instance	**time**	$\overline{c(T)}$	σ	$\overline{c(T)}$	σ	$\overline{c(T)}$	σ	$\overline{c(T)}$	σ	$\overline{c(T)}$	σ
kroa150	150s	9830.6	31.4	9831.3	30.1	9815.0	0.0	9815.0	0.0	9815.0	0.0
d198	300s	7055.1	8.7	7059.6	9.0	7044.6	2.3	7044.0	0.0	7044.0	0.0
krob200	300s	11275.0	45.6	11248.9	7.5	11244.0	0.0	11244.0	0.0	11244.0	0.0
gr202	300s	242.1	0.3	242.2	0.4	242.0	0.2	242.0	0.0	242.0	0.0
ts225	300s	62290.8	40.4	62299.1	50.9	62268.6	0.5	62268.4	0.5	62268.3	0.5
pr226	300s	55515.0	0.0	55515.0	0.0	55515.0	0.0	55515.0	0.0	55515.0	0.0
gil262	450s	945.5	4.0	945.0	3.7	942.4	2.0	942.0	0.0	942.0	0.0
pr264	450s	21893.2	7.7	21898.4	20.9	21886.0	0.0	21886.0	0.0	21886.0	0.0
pr299	450s	20352.1	37.4	20349.7	24.9	20318.5	11.3	20318.1	11.3	20316.0	0.0
lin318	600s	18545.9	29.2	18547.3	25.6	18525.8	12.4	18511.0	10.8	18513.8	9.9
rd400	600s	5953.0	15.4	5959.4	20.2	5946.4	10.8	5940.2	6.5	5939.7	6.7
fl417	600s	7982.0	0.0	7982.0	0.0	7982.0	0.0	7982.0	0.0	7982.0	0.0
gr431	600s	1034.1	1.4	1033.4	0.9	1033.3	0.7	1033.0	0.0	1033.0	0.0
pr439	600s	51921.4	60.7	51888.5	56.3	51810.5	26.5	51791.0	0.0	51791.0	0.0
pcb442	600s	19717.0	59.5	19708.1	70.2	19632.6	21.1	19623.7	15.9	19617.0	12.4

Table 6.6 compares the average qualities of the obtained final solutions for the following EA variants when terminating after the same CPU-time limits: the EA without any archive, the EA with the NSB archive, the EA with the GESB archive, the EA with both archives, and the EA with both archives and bounding. The first two columns list the instance names and the time limit. For each EA variant we show the average final objective values $\overline{c(T)}$ and corresponding standard deviations (σ). Best results are highlighted. We observe that the EA variant without archive performs worst in general. Among the two variants where the archive only uses one representation, the GESB representation is more often the better choice. By combining both archives we get significantly better results, and the best results except on a single instance are obtained when additionally applying bounding. This indicates that the solution archives clearly have a positive impact on the EA, and the time overhead introduced by building and maintaining the trie data structure(s) pays off. Differences between the two variants with and without the bounding extension are small, however. A one-sided Wilcoxon rank sum test with an error level of 5% did not indicate the significance of the observed difference in the average values.

To obtain a more detailed picture, we further compare in Table 6.7 the last two EA variants using both archives with and without bounding on an extended TSPlib-based instance set introduced by Öncan et al. [209][3]. These instances are larger and were also derived by geographical center clustering. Note that some of these instances have the same names as those in the first set, but the clustering data and/or the distance data are different. Beside the average final objective values and corresponding standard deviations, we also list for each instance the error probability p of a one-sided Wilcoxon rank sum test for the assumption that the variant with bounding performs better than the variant without bounding. Apart from five instances where both variants perform equally well and one instance where the EA variant without bounding performs slightly better, we observe that bounding con-

[3] http://neumann.hec.ca/chairedistributique/data/gmstp

Table 6.7 Results of the EA using both solution archives with and without bounding, respectively, for the extended TSPlib-based instances from Öncan et al. [209]

		no bounding		with bounding		
instance	**time**	$\overline{c(T)}$	σ	$\overline{c(T)}$	σ	p
ali535	600s	114581.1	95.8	114404.9	90.7	<0.001
att532	600s	12008.1	7.4	12004.9	4.5	0.051
d493	600s	16516,8	14,5	16494.9	3.1	<0.001
d657	600s	19504.0	42.2	19451.7	31.3	<0.001
fl417	600s	7935.0	0.0	7935.0	0.0	N.A.
gil262	450s	887.0	0.0	887.0	0.0	N.A.
gr431	600s	86889.2	18.0	86888.2	17.3	0.500
gr666	600s	144837.3	109.1	144790.6	77.1	0.037
lin318	600s	18485.9	13.9	18485.1	8.6	0.468
p654	600s	22207.0	0.0	22207.0	0.0	N.A.
pa561	600s	870.6	2.9	866.7	2.4	<0.001
pcb442	600s	19589.0	21.5	19584.1	15.0	0.109
pr264	450s	21872.0	0.0	21872.0	0.0	N.A.
pr299	450s	20301.2	16.3	20290.0	0.0	<0.001
pr439	600s	51754.3	15.1	51749.0	0.0	0.050
rat575	600s	2184.0	6.3	2178.5	4.8	<0.001
rat783	600s	3033.6	13.2	3028.5	8.3	0.076
rd400	600s	5874.9	11.2	5875.0	5.4	0.670
si535	600s	12791.0	0.0	12791.0	0.0	N.A.
u574	600s	15063.0	15.1	15051.4	8.9	<0.001
u724	600s	15965.1	35.4	15949.2	29.4	0.046

sistently increases the solution quality. For seven instances these improvements are statistically significant. Thus, these tests back up the assumption that the bounding extension is beneficial for the solution archive in this context.

Table 6.8 further compares the EA using both archives and bounding with other state-of-the-art approaches from the literature including the tabu search approach (TS1) by Ghosh [109], the hybrid variable neighborhood search with the global subtree optimization neighborhood (COMB*-VNS) by Hu et al. [142] detailed in Section 4.3, and an algorithm based on dynamic candidate sets (DCS) by Jiang and Chen [146]. It can be seen that the archive-enhanced EA competes well with those other approaches, especially on larger instances. We remark, however, that the runtimes of TS1 and DCS are difficult to compare as these algorithms were tested on different machines; results are adopted from the corresponding publications.

Table 6.9 further compares the archive-enhanced EA with the genetic algorithm (GA) by Golden et al. [117] and the tabu search approach (TS2) by Öncan et al. [209] on the extended TSPlib-based instances. Since these authors did not use a fixed CPU-time as termination criterion, we terminated our EA after 2000 iterations without improvement. Because only best solutions are listed for GA and TS2 in [209], we also focused on the best solutions obtained by 30 runs for the archive-enhanced EA, but nevertheless also list average values and the standard deviations. We observe that the archive-enhanced EA competes well with TS2 and clearly outperforms the GA. For instance fl417 the EA consumes an exceptionally large

Table 6.8 Comparison of the EA utilizing both solution archives and bounding with other state-of-the-art approaches for the standard TSPlib-based instances

		TS1	COMB*-VNS		DCS		EA+arch.+bound.	
instance	time	$c(T)$	$\overline{c(T)}$	σ	$\overline{c(T)}$	σ	$\overline{c(T)}$	σ
kroa150	150s	9815.0	9815.0	0.0	9815.0	0.0	9815.0	0.0
d198	300s	7062.0	7044.0	0.0	7044.0	0.0	7044.0	0.0
krob200	300s	11245.0	11244.0	0.0	11244.0	0.0	11244.0	0.0
gr202	300s	242.0	242.0	0.0	242.0	0.0	242.0	0.0
ts225	300s	62366.0	62268.5	0.5	62268.3	0.5	62268.3	0.5
pr226	300s	55515.0	55515.0	0.0	55515.0	0.0	55515.0	0.0
gil262	450s	942.0	942.3	1.0	942.0	0.0	942.0	0.0
pr264	450s	21886.0	21886.5	1.8	21886.0	0.0	21886.0	0.0
pr299	450s	20339.0	20322.6	14.7	20317.4	1.5	20316.0	0.0
lin318	600s	18521.0	18506.8	11.6	18513.6	7.8	18513.8	9.9
rd400	600s	5943.0	5943.6	9.7	5941.5	9.9	5939.7	6.7
fl417	600s	7990.0	7982.0	0.0	7982.7	0.5	7982.0	0.0
gr431	600s	1034.0	1033.0	0.2	1033.0	0.0	1033.0	0.0
pr439	600s	51852.0	51847.9	40.9	51833.8	36.1	51791.0	0.0
pcb442	600s	19621.0	19702.8	52.1	19662.5	39.8	19617.0	12.4

Table 6.9 Comparison of the EA utilizing both solution archives and bounding with other state-of-the-art approaches for the extended TSPlib-based instances

	GA		TS2		EA+arch.+bound.			
instance	$c^*(T)$	time	$c^*(T)$	time	$c^*(T)$	$\overline{c(T)}$	σ	time
ali535	114379	492s	114303	683s	114303	114419.1	96.6	243s
att532	12007	500s	12001	597s	12001	12007.8	6.3	115s
d493	16526	388s	16493	587s	16493	16501.0	18.3	154s
d657	19465	969s	19427	1056s	19427	19456.6	32.5	335s
fl417	7936	218s	7935	233s	7935	7935.0	0.0	2570s
gil262	887	73s	887	74s	887	887.0	0.0	13s
gr431	86899	266s	86885	233s	86885	86903.4	42.3	80s
gr666	144918	2866s	144756	1365s	144737	144747.7	40.8	237s
lin318	18476	105s	18471	130s	18471	18486.3	4.3	46s
p654	22214	881s	22208	1045s	22207	22207.0	0.0	1634s
pa561	868	559s	864	702s	865	870.7	2.7	107s
pcb442	19670	284s	19571	266s	19571	19593.9	22.1	66s
pr264	21886	57s	21872	72s	21872	21872.0	0.0	20s
pr299	20307	86s	20290	94s	20290	20290.0	0.0	21s
pr439	51808	981s	51760	574s	51749	51749.9	3.6	84s
rat575	2189	627s	2170	762s	2170	2180.0	6.3	125s
rat783	3044	1653s	3017	1916s	3015	3027.4	7.4	292s
rd400	5880	205s	5868	208s	5868	5875.9	8.9	56s
si535	12791	458s	12791	573s	12791	12791.0	0.0	123s
u574	15069	620s	15037	517s	15034	15058.1	14.3	155s
u724	16015	1281s	15905	1290s	15904	15947.4	26.6	324s

amount of runtime. The reason is that the final best solution can be found extremely easily in a few seconds by the EA, but with its solution archives it then spends much time on generating and transforming a large number of duplicates.

Table 6.10 Average time and memory consumptions of the EA with either the NSB or the GESB archive with and without bounding

	EA with NSB archive				EA with GESB archive			
	no bounding		with bounding		no bounding		with bounding	
instance	time	mem	time	mem	time	mem	time	mem
kroa150	23s	6.7 MB	36s	8.0 MB	24s	28.9 MB	37s	32.8 MB
d198	39s	10.7 MB	75s	12.5 MB	41s	55.4 MB	58s	59.5 MB
krob200	49s	8.2 MB	82s	9.7 MB	49s	49.0 MB	99s	54.1 MB
gr202	42s	10.0 MB	68s	11.6 MB	44s	58.3 MB	55s	61.2 MB
ts225	49s	17.3 MB	73s	19.0 MB	53s	84.7 MB	61s	85.3 MB
pr226	64s	6.8 MB	126s	7.7 MB	61s	76.1 MB	100s	89.3 MB
gil262	74s	17.9 MB	123s	21.3 MB	78s	111.4 MB	93s	115.8 MB
pr264	80s	15.1 MB	139s	17.8 MB	83s	108.5 MB	106s	115.9 MB
pr299	101s	18.8 MB	174s	22.1 MB	105s	133.6 MB	129s	137.4 MB
lin318	113s	17.8 MB	217s	22.1 MB	116s	163.5 MB	152s	174.2 MB
rd400	168s	25.8 MB	318s	31.9 MB	178s	264.7 MB	212s	276.9 MB
fl417	204s	18.1 MB	499s	20.5 MB	196s	265.0 MB	326s	302.5 MB
gr431	243s	28.9 MB	466s	34.3 MB	248s	309.3 MB	267s	315.3 MB
pcb442	217s	34.4 MB	406s	41.1 MB	229s	360.0 MB	244s	367.4 MB

Table 6.10 finally shows the impact of applying either the NSB archive or the GESB archive with and without bounding on the EA's time and memory consumption when performing 10000 iterations. Applying only one of the archives allows us to see their individual demands more clearly. For each variant we show average CPU-times and the average sizes of the archives upon termination after the 10000 iterations. As already argued in Section 6.3.1, the GESB trie is in general substantially larger than the NSB trie, although the overall memory consumption is in all cases reasonable—400 MB were never exceeded. Even with the low probability of 5% per node, the time overhead of the bounding is substantial and in many cases the runtime was doubled. We also observe a slightly increased memory consumption when applying bounding. This might be counterintuitive at first glance since one goal of this technique is to save memory by pruning subtries. However, a new solution is generated by the transform operation each time a subtrie is pruned, frequently introducing a series of new nodes. Thus, by an effective pruning, larger parts of the search space are covered in the limited number of iterations. When applying both solution archives in the EA together, runtimes and memory requirements were almost always less than the sum of the listed values for the individually applied solution archives.

6.4 Other Applications of the Concept of Complete Solution Archives

Besides the related work already summarized in Section 6.1.1, the specific idea of the complete solution archive presented here were applied in several other works.

Ruthmair and Raidl [263] describe a memetic algorithm for the rooted delay-constrained minimum spanning tree problem, in which a graph with costs and delays associated to the edges is given and a minimum cost spanning tree is sought under the constraint that the path from a dedicated root to each node may not have a larger delay than a certain bound B. The used trie is structured differently than in our examples for the GMST problem: Assuming not too large integer values for the delay bound B, a candidate solution is indirectly represented by the vector of total delays for each node except the root, and a decoder that connects each node to the least-cost node at a preceding delay-level is used for deriving the actual tree. Each trie node consequently contains an array of B references to nodes at the next level, and at each level a dedicated node's delay in a given solution array decides which pointer to follow. Some special adaptions are applied to the trie in order to reduce the used space. In particular, not all delay values are feasible for a node, so the number of array elements of a trie node can be reduced accordingly. Experimental results indicate that this archive may be beneficial for instances with low delay bounds or if the number of revisits is very high, but a simpler hash-based archive also works well in the context of the considered memetic algorithm and has a smaller overhead.

Zaubzer [303] applied a trie-based solution archive to a memetic algorithm for the multidimensional knapsack problem. Here, the knapsack constraints make the situation more complicated. Substantial problem-specific extensions have thus been considered for the archive in order to only produce candidate solutions on the boundary of the feasible region. Depending on the test instances and configuration, some benefits could be observed when using this archive. However, due to the strong repair and local improvement procedures embedded in this memetic algorithm, general advantages turned out to be rather small. Nevertheless, this work also gives a clear indication that the basic idea of enhancing an EA by a trie-based complete solution archive is promising.

Biesinger et al. [18] enhanced a genetic algorithm for reconstructing cross-cut shredded text documents by a complete solution archive. Candidate solutions are represented by a special vector encoding the snippets' positions, and the trie is realized as a linked trie, in which nodes are stored as linked lists for saving space. Experimental results indicate that the solution archive is able to improve the performance of the GA w.r.t. solution quality.

Last but not least, Biesinger et al. [20] successfully applied a complete solution archive in a hybrid genetic algorithm for the Σ_2^P-hard discrete $(r|p)$-centroid problem. In this competitive facility location problem, two non-cooperating players enter a market sequentially, and it is the first player's goal to place its facilities in an optimal way assuming an optimal response of the second player. The trie-based solution archive stores the locations for the first player's facilities, effectively avoiding the re-consideration of the same configurations and thus resolving the corresponding subproblems of the second player. Besides the solution archive, the genetic algorithm is enhanced by an embedded tabu search acting as advanced local improvement and a multilevel solution evaluation scheme. The tabu search also uses the solution archive as its tabu memory. Together, these features make the proposed approach one of the currently leading solution approaches for the considered problem.

This approach was further successfully enhanced in [19] to leader-follower facility location problems with proportional customer behavior, and in [21] to binary behavior in the variants with essential and inessential demand. Again, the employed solution archive provided here a substantial basis for the success.

More generally, the concept of complete solution archives as presented in this chapter is clearly not one that can be meaningfully applied as a general purpose tool in any EA or metaheuristic. It may make much sense, however, in cases where the representation of solutions is rather compact, the evaluation of individual solutions is expensive, and the chance to reconsider identical solutions would be rather high—possibly due to a strong intensification in the metaheuristic search. The aspect of expensive evaluations includes in particular problems where solution candidates must undergo some costly (in terms of running times but possibly also other aspects) simulation or problems where candidate solutions are indirectly or incompletely represented and a complex method is applied to derive the actual solution. The solution archive's underlying data structure must be carefully designed: Most importantly, the operations of searching/checking a solution, inserting a new one, and transforming a duplicate into a usually similar, guaranteed new solution must be realized in highly efficient ways. Different variants of trie data structures, as described here, allow this in the considered cases. However, we could also observe that this data structure in general needs to be adapted to the specifically used representation. Furthermore, constraints making only part of the search space feasible may impose an additional challenge, as we have seen in the case of the GESB trie for storing spanning trees encoded by their predecessor vector. In particular coming up with an effective transform function may then be a challenge that needs to be solved in a highly problem-specific way.

Chapter 7
Further Hybrids and Conclusions

In this final chapter we summarize further important strategies of hybrid metaheuristics that have been successfully applied in a variety of applications, and conclude with some final thoughts on this research field. In particular, we elaborate on the question of under which conditions the consideration of hybrid metaheuristics might be meaningful in practical scenarios.

7.1 Combinations of Metaheuristics with Other Heuristic Methods

Historically the first way to combine metaheuristics with other techniques for optimization concerned the combination with other heuristics and metaheuristics. In fact, the most typical hybridization in this context makes use of local search methods—or even metaheuristics based on local search—within population-based techniques. In particular, local search methods are used in metaheuristics such as evolutionary algorithms and ant colony optimization for improving the solutions generated by these techniques. This type of hybridization—which, in the context of evolutionary algorithms, is known as the *Memetic Algorithm* [201]—is very successful, mainly for the following reason. In general it is said that population-based metaheuristics excel in identifying areas of the search space containing high quality solutions. However, they are not as good at finding the best solutions in these areas. In contrast, local search methods—and, more generally, metaheuristics based on local search—are typically good at finding the best or excellent solutions in a confined area of the search space, while identifying these areas is not among their strengths. Therefore, by combining both types of techniques the algorithm designer profits from their complementary strengths.

However, in recent years some hybrid metaheuristics have emerged in which concepts of population-based methods are beneficially used within the framework of metaheuristics based on local search. Prominent examples are algorithms such as *Population-based Iterated Local Search* [276] and *Population-based Iterated*

Greedy Algorithms [42]. The main idea of these techniques is quite simple. Instead of working on a single incumbent solution, both population-based ILS and IG maintain at all times a population P of n solutions. The usual algorithmic operators that are applied to the incumbent solution at each iteration are, instead, applied to each solution $s \in P$. Adding all solutions generated in this way to P results in a population P' of $2n$ solutions. Generally, the population for the next generation of the algorithm is obtained by choosing the best n solutions from P'. Another example for using a population-based technique within a metaheuristic based on local search can be found in [182], where the authors use an evolutionary algorithm as a perturbation technique for iterated local search.

Multilevel Techniques [294, 295] are heuristic frameworks developed for dealing with large-scale problem instances. They are based on the following idea. Starting from the original problem instance, smaller and smaller instances are derived by successive *coarsening* until some stopping criteria are met. Coarsening, for example, might unite entities of the problem by joining them. In this way, a hierarchy of problem instances is generated, where the largest one corresponds to the original problem instance. In general, the problem instance corresponding to a certain level is always smaller than (or of equal size as) the problem instance corresponding to the next higher level. Next, an optimization technique such as, for example, a metaheuristic is used to generate a good-enough solution to the smallest problem instance in this hierarchy. This solution is successively expanded into a solution to the problem instance corresponding to the next level until a solution to the original problem instance is obtained. Additionally, after being expanded to each next level, solutions may be subject to a refinement process. For example, the same metaheuristic used to generate a solution for the smallest problem instance might be used for this purpose. Occasionally, the coarsening and refinement phases are also iterated. Among the earliest applications were those to mesh partitioning [296], the traveling salesman problem [293], and graph coloring [294].

The principal idea behind *Variable Fixing* strategies is related to that of multilevel techniques. More specifically, variable fixing refers to choosing fixed values for a subset of the given problem's variables. Optimization is then only performed over the rest of the search space. A first example concerns the so-called *Core Concepts* in the context of knapsack problems [229, 241]. Another example where variable fixing is essential is the variable neighborhood decomposition approach proposed in [170]. Finally, *Problem Kernelization*—which is a systematic approach based on tools from the field of parameterized complexity—is also related to multilevel strategies and variable fixing. The basic procedure consists in reducing a given problem instance in polynomial time to its so-called problem kernel. An optimal solution to the problem kernel can then be transformed in polynomial time to an optimal solution to the original problem instance. In [110], Gilmour and Dras proposed an ant colony optimization algorithm that makes use—in several different ways—of the above mentioned problem kernels. Another idea related to variable fixing is motivated by so-called *Backbones* of optimization problems. The backbone of an optimization problem is defined as the set of solution components (variable-value assignments) which must form part of an optimal solution [272]. In

Backbone-guided Search, approximate backbones are used to guide the search of a metaheuristic [305].

7.2 Hybridization by Efficiently Solving Problem Relaxations

As already mentioned in Section 1.5 in the context of mixed integer programming, *problem relaxations* frequently provide an important basis for obtaining effective optimization algorithms. By dropping or weakening certain constraints of an original problem formulation, a problem may become efficiently solvable. On the one hand, the solution to such a relaxation provides a bound for the optimal solution to the original problem; remember that MIP solvers applying LP-based branch-and-bound in their core make rigorous use of such dual bounds obtained from the LP relaxation. On the other hand, the solution to the relaxed problem frequently also provides a promising starting point for heuristics and metaheuristics. For example, solutions to LP relaxations are frequently rounded or in some other way repaired to obtain a feasible integral solution in their proximity, variables that already have integral values in the LP solution might be fixed, or LP variable values may be used to bias variation operators, e.g., by choosing close variable values with higher probabilities.

Occasionally, dual variable information of LP solutions may provide even more helpful guidance. Chu and Beasley [57], for example, make use of this in a genetic algorithm for the multidimensional knapsack problem by calculating so-called *pseudo-utility ratios* for the primal variables and steering with them repair and local improvement strategies. Also Puchinger et al. [242] point out that at least for the multidimensional knapsack problem pseudo-utility ratios are significantly better indicators for the likelihood of the corresponding items appearing in an optimal integer solution than primal LP variable values.

Apart from the LP relaxation, other relaxations are sometimes exploited in conjunction with metaheuristics. Besides problem-specific approaches, Lagrangian relaxation [99] has been particularly successful [130, 145, 171, 223, 284]. Lagrangian relaxation drops one or more constraints and instead adds respective penalty terms weighted by so-called Lagrange multipliers to the objective function. These multipliers are then optimized, e.g., by some subgradient method, in order to obtain a best achievable bound. In comparison to the LP relaxation, Lagrangian relaxation has the advantage of frequently yielding tighter bounds. This, however, comes at the cost of a usually higher computational effort.

7.3 Strategic Guidance of Branch-and-Bound by Metaheuristics

As already mentioned when outlining the principle of branch-and-bound in Section 1.3, primal as well as dual bounds play a crucial role for branch-and-bound.

Primal bounds are typically delivered from feasible solutions, and thus effective heuristics—including fast metaheuristics—are frequently used for finding good initial solutions as well as new incumbent solutions to subproblems within the branch-and-bound process.

Besides these most natural applications of (meta)heuristics within branch-and-bound, principles of local-search-based metaheuristics are sometimes mimicked by special control strategies for selecting the tree nodes (i.e., open subproblems) to be processed next or by special branching strategies focusing the tree search to neighborhoods of promising solutions.

Danna et al. [72] proposed *Guided Dives*, where the tree search temporarily switches to a depth-first strategy and always considers next a subproblem where the branching variable has the value of the incumbent solution. Guided dives are repeatedly applied in regular intervals. The results indicate a strongly improved heuristic performance.

On the contrary, Fischetti and Lodi [98] suggested *Local Branching*. Given an incumbent solution again, branching is performed by adding on the one hand a so-called local branching constraint that restricts the search space to the incumbent's k-opt neighborhood and on the other hand its inverse representing the remaining search space. The MIP solver is then forced to completely solve the k-opt neighborhood before considering the remaining open nodes of the branch-and-bound tree. If an improved solution has been found, a new subproblem corresponding to the k-opt neighborhood of the new incumbent is split off, otherwise a larger k may be tried. When no further improvements are achieved, the remaining problem is processed in a standard way. While local branching is often beneficial, the authors found that the addition of the inverse local branching constraints frequently is counterproductive as many of these dense constraints degrade performance. Note that local branching can also be interpreted as a kind of large neighborhood search, as we already observed in Section 4.4 and where we discussed the local branching constraint in detail.

Danna et al. [72] further proposed *Relaxation Induced Neighborhood Search* (RINS), where occasionally a sub-MIP is spawned from a search-tree node corresponding to another special neighborhood of an incumbent solution: Variables having the same values in the incumbent and the current solution to the LP relaxation are fixed and an objective value cutoff corresponding to the current LP value is set. The subproblem on the remaining variables is then solved with limited time. In the authors' experimental comparison of guided dives, local branching, and RINS on a collection of MIP models originating from diverse applications, RINS performed best. It has been included in CPLEX as a standard strategy for boosting heuristic performance.

Recently, Gomes et al. [120] suggested an extension of RINS that explicitly explores pre-processing techniques. Their method systematically searches for a suitable number of fixations to produce subproblems of controlled size, which are explored in a variable-neighborhood-descent fashion.

Last but not least, note that the complete solution archive approach presented in Chapter 6 also can be seen as a variant where a tree search is guided by metaheuristic principles. The relation to branch-and-bound becomes especially clear when considering the bounding extension discussed in Sections 6.3.4 and 6.3.5.

7.4 Mathematical Programming Decomposition Based Hybrids

In the area of mathematical programming, it has been recognized that practical problems frequently have a special structure that can be exploited by reformulating and decomposing an original model into two or more dependent smaller problems. Certain techniques then allow us to solve the original problem more efficiently. Three such decomposition techniques are particularly well known: Lagrangian decomposition, Dantzig–Wolfe decomposition in conjunction with column generation, and Benders decomposition. It has been recognized that these concepts also provide a very fruitful basis for effective hybrid metaheuristics.

We briefly summarize the key ideas and hybridization possibilities of these three decomposition techniques in the following subsections. For more detailed information on such metaheuristic hybrids see the articles from Boschetti and Maniezzo [40, 41] and the recent survey from Raidl [244].

7.4.1 Lagrangian Decomposition

Lagrangian Decomposition (LD) is based on the principle of Lagrangian relaxation sketched out earlier in Section 7.2. The relaxation is done, however, in a way such that the original problem decomposes into two or more subproblems that have disjoint sets of variables and constraints and are only connected by the objective function. In the course of solving the Lagrangian dual, i.e., finding the optimal Lagrangian multipliers, the subproblems can thus be individually solved in typically much more efficient ways.

For example, Pirkwieser et al. [228] approach the knapsack constrained maximum spanning tree problem by duplicating the decision variables for the edges and relaxing the linking constraints in a Lagrangian fashion. The obtained subproblems correspond to the classic 0–1 knapsack problem and the maximum spanning tree problem, which can both be solved efficiently in polynomial time. In their hybrid metaheuristic, they first solve this LD and then utilize its intermediate and final results within an evolutionary algorithm. On the one hand, the original problem is reduced by only considering edges appearing in the LD's intermediate results. On the other hand, reduced costs obtained from the Lagrangian multipliers are exploited by heuristically guiding the EA's variation operators. Remarkably, the approach was able to solve many very large benchmark instances to proven optimality. Optimality could in these cases be recognized, as the bounds obtained by the LD coincided with the solution's objective values.

Similar approaches have, for example, been described by Haouari and Siala for the Steiner tree problem [130] and by Leitner and Raidl [171] for a network design problem concerning the last mile in fiber optic networks.

7.4.2 Dantzig–Wolfe Decomposition and Column Generation

Dantzig–Wolfe Decomposition [73] was originally introduced for solving very large LPs having a special block-diagonal structure and relies on *(Delayed) Column Generation* (CG). In mathematical programming, CG is a well-known technique to approach in particular MIP models involving a huge, possibly exponentially large, number of variables. Such MIPs frequently arise in set-cover or set-partitioning formulations, e.g., in vehicle routing, network design, cutting and packing, and scheduling problems. Often, such models can be shown to be substantially stronger than alternative compact formulations w.r.t. their LP relaxations. In these models, variables represent, for example, whole routes, paths, or (sub)patterns in packing and scheduling problems.

In principle, CG works by starting with a very small subset of the huge set of variables and solving the respective LP relaxation, called the *restricted master problem*. The dual variable values of the obtained solution then give rise to the *pricing subproblem*, which when solved yields further variables (i.e., columns in the LP's matrix notation) whose inclusion in the restricted master problem in turn may lead to an improved master solution. The augmented restricted master problem is thus resolved, and the whole process iterated until no further advantageous variables exist. The LP relaxation of the original problem has then been solved to optimality without considering all original variables explicitly. To ultimately obtain a proven optimal solution to the original MIP, CG needs to be embedded in a branch-and-bound. This is then called *branch-and-price*. For in-depth information on CG we refer to [76] and [183].

In a heuristic context CG provides a highly fruitful basis for hybridization:

- Obviously, the frequently excellent solution to the LP relaxation obtained from CG can be exploited in the ways already sketched above.
- Furthermore, researchers have proposed to determine the initial variables of the restricted master problem from diverse good heuristic solutions obtained, for example, from a population-based metaheuristic. CG for a set-partitioning or set-covering model can then be also interpreted as a kind of optimal merging, see Section 3.3.2.
- Also, (meta)heuristics can be highly useful for solving difficult pricing subproblems in faster ways, see, for example, [240].
- Last but not least, the solution components corresponding to variables that are priced in the course of CG can provide highly meaningful building blocks for metaheuristics, especially together with their LP values and possibly dual variable values associated with the model's constraints.

For successful examples of the above mentioned concepts, see Filho et al. [254], who describe a constructive genetic algorithm in conjunction with CG for graph coloring; Pirkwieser and Raidl [226], who consider a variable neighborhood search and an evolutionary algorithm collaborating with CG to solve the periodic vehicle routing problem with time windows; and Massen et al. [190], who apply ant colony optimization for heuristic CG to solve a black-box vehicle routing problem. Massen et al. [191] show how the latter pheromone-based heuristic CG can further be improved by automatic algorithm configuration methods.

Alvelos et al. [4] describe a general hybrid strategy called SEARCHCOL, where CG and a metaheuristic are iteratively performed and information is exchanged between both. The metaheuristic works in a problem-independent way trying to find a best integral solution by searching over combinations of variables identified in CG, while the CG is perturbed in each iteration based on the metaheuristic's result by fixing subproblem variables with special constraints.

7.4.3 Benders Decomposition

Benders Decomposition (BD) was originally suggested for solving large MIPs composed of subproblems having their own variables and constraints but also involving "complicating" variables that connect these subproblems [15]. The idea is to define a master problem only on the connecting variables and consider the contributions of the subproblems by iteratively adding respective inequalities. Benders decomposition can be regarded dual to CG, as instead of iteratively adding variables (i.e., columns) to a master problem, inequalities (i.e., rows) are added. In contrast to LD and CG, BD is able to directly yield an optimal solution to the original MIP, not only a relaxation of it.

While classic BD has the major restriction that the subproblems must be LPs, thus they may in particular only contain continuous variables, generalizations like the *Logic-based BD* [136] or *Combinatorial Benders Cuts* [61] exist, allowing also integral variables and nonlinear relationships.

As with LD and CG, there are several possibilities for metaheuristics to come into play with BD. Although smaller than the original problem, the master problem is frequently still difficult to solve, especially when many Benders cuts have been added. Metaheuristics may provide good approximate master problem solutions in much shorter time, and these solutions may be sufficient in order to achieve substantial speedups of the overall approach. If, finally, when the metaheuristic cannot identify an improved master solution, the master problem is solved to optimality and no further Benders cuts can be identified, the whole approach is still complete. Poojari and Beasley [231] describe such an approach for solving general MIPs in which a genetic algorithm together with a feasibility pump heuristic are applied to the master problem. The authors argue that a population-based metaheuristic like a genetic algorithm is particularly useful as it provides multiple solutions in each iteration giving rise to more Benders cuts. Similarly in spirit, Lai et al. [168] propose a genetic algorithm/BD hybrid for solving the capacitated plant location problem; results indicate a tremendous saving of computation time in comparison to classic BD. Lai et al. [169] further discuss such an approach for a vehicle routing problem. Rei et al. [252] suggest to use local branching, see Section 7.3, for solving a MIP master problem in order to sooner find improved upper as well as lower bounds.

Initial solutions obtained by some (meta)heuristic may be used to derive an initial set of Benders cuts in order to start with a more meaningful first master problem. For example, Easwaran and Üster [85] apply a tabu search to warm-start a BD approach for a supply chain network design problem.

Subproblems need not always be solved to optimality in order to obtain useful Benders cuts, even when completeness of the whole approach shall be retained [302]. Especially when considering difficult subproblems in logic-based BD, constraint programming and metaheuristics have great potential for speeding up the overall approach by providing helpful cuts much faster. For example in [135] a constraint-programming-based BD is described that substantially outperforms pure MIP as well as constraint programming approaches on a large class of planning and scheduling problems. Cordeau et al. [64] solve an aircraft routing and crew scheduling problem by applying BD and use CG-based heuristics for the master problem as well as the integer subproblem.

Finally, Raidl et al. [248] proposed an exact logic-based BD approach for a bi-level capacitated vehicle routing problem and sped it up considerably by first solving all instances of the master problem as well as all subproblems by means of a fast variable neighborhood search heuristic. Invalid Benders cuts possibly cutting off feasible solutions might be created. In a second phase, all these heuristically generated Benders cuts are verified by re-solving the corresponding subproblems exactly by means of the MIP solver, yielding possibly corrected cuts that replace the invalid ones. When finally the master problem is solved exactly and no further Benders cuts can be derived, a proven optimal solution is obtained.

7.5 Conclusions

With this book we have made an attempt to present and summarize from the area of hybrid metaheuristics, on the one side, some of the most popular techniques and, on the other side, some of the rather recent but highly promising techniques. In particular, we have chosen to discuss—and illustrate by means of examples—five general hybridization techniques based on different ideas:

- *Indirect solution representations and decoders:* The main idea here is to choose a compact solution representation which permits the use of rather standard search operators. Moreover, the representation should transfer some aspects of the problem difficulty to the decoder algorithm that serves as a mapping between the representation and actual solutions to the problem.

- *Instance size reduction:* In those cases in which it is possible to substantially reduce the size of a problem instance—for example, by removing solution components—in such a way that the reduced problem instance still contains high-quality solutions to the original problem instance, it might be feasible to solve the reduced problem instance even to optimality by applying appropriate exact techniques.
- *Large neighborhood search:* This is one of the most generally applicable algorithms from the field of hybrid metaheuristics, based on the simple idea of defining large neighborhoods that can be efficiently explored, mostly by exact techniques such as general MIP solvers, dynamic programming, or constraint programming.
- *Parallel and non-independent construction of solutions:* The main idea is the incorporation of tree search concepts such as the use of bounding information (in addition to greedy information) and the parallel and non-independent construction of solutions—such as in beam search—into metaheuristics.
- *Complete solution archives:* Based on the observation that metaheuristics sometimes waste a substantial amount of computation time by allowing us to revisit solutions during a run, the main idea is to avoid the repeated generation of solutions by means of intelligent data structures that serve as complete solution archives. These archives are in particular able to efficiently guide the search to so-far unvisited solutions similar to already generated ones, and might further be extended by pruning larger parts of the search space via bounding.

Apart from these five main ideas, the last chapter of this book summarizes hybrid approaches based on the combination of different (meta)heuristics, problem relaxations, the strategic guidance of branch-and-bound methods by means of metaheuristics, and mathematical programming decomposition techniques. We hope that this book will serve as a useful starting point for researchers and practitioners who want to learn about hybrid metaheuristics.

Remember that the main focus of practical optimization should always be on solving the problem at hand in the best, and most efficient, way possible. Needless to say this goal cannot always be achieved by means of hybrid metaheuristics. Depending on the characteristics of the problem—and the specific problem instances—to be solved, it might be perfectly possible to use, for example, a general purpose MIP solver, or a simple—in the sense of pure—metaheuristic. In other words, one should not try to crack a walnut with a sledgehammer. Moreover, the process of designing and implementing effective hybrid metaheuristics can sometimes be rather complicated and may involve knowledge and experience in a broad spectrum of algorithmic techniques, programming and data structures, as well as algorithm engineering and statistics. Nevertheless, answering the following questions may help in deciding whether to start the adventure of developing a hybrid metaheuristic:

1. **What is the optimization goal?** Is a reasonably good solution required in a very short amount of computation time, or does it pay to spend more implementation

and computation time in order to produce (near-)optimal solutions? In particular, in those cases in which (near-)optimal solutions are needed for problem instances too large to be tackled by the available exact techniques and too difficult to be solved by pure metaheuristics, the development of a hybrid metaheuristic seems appropriate.

2. **Is there room for improvement over the current state-of-the-art techniques?** It might be the case that the existing pure metaheuristic strategies work very well for the problem instances under consideration. Alternatively, the problem instances under consideration might already be solvable by complete techniques in a reasonable amount of computation time. In these cases, spending time and effort on developing hybrid metaheuristics might be in vain.

3. **Which type of hybrid metaheuristic might work well for the problem at hand?** Unfortunately, even when only considering pure metaheuristics, it is hardly possible to provide a well-founded answer to this question. In the case of hybrid metaheuristics this is even more difficult. In other words, it is hardly possible to find general guidelines. In the five main chapters of this book we have identified five generally applicable techniques with very different characteristics that make them more or less suited for certain optimization problems. According to our experience and professional judgement, large neighborhood search, for example, seems rather appropriate for problems in which the number of solution components is linear in the input parameters. In contrast, algorithms based on problem instance reduction appear to work better in the context of optimization problems with a super-linear number of solution components. However, statements of this kind should be regarded with caution, because currently they are solely based on experience and they lack any kind of theoretical foundation. In general, we recommend (1) a careful literature search with the aim of identifying the most successful optimization approaches for the problem at hand or for similar problems, and (2) the study of different ways of combining the most promising features of the identified approaches.

Research on hybrid metaheuristics is gradually gaining maturity. In the coming years, many research efforts concerned with approximate problem solving will be dedicated to hybrid techniques. We hope that this book contributes by giving some structure and guidance to this interesting line of research.

References

[1] Aggarwal C, Orlin J, Tai R (1997) Optimized crossover for the independent set problem. Operations Research 45:226–234

[2] Ahuja R, Orlin J, Tiwari A (2000) A greedy genetic algorithm for the quadratic assignment problem. Computers & Operations Research 27:917–934

[3] Akeb H, Hifi M, M'Hallah R (2009) A beam search algorithm for the circular packing problem. Computers & Operations Research 36(5):1513–1528

[4] Alvelos F, de Sousa A, Santos D (2013) Combining column generation and metaheuristics. In: Talbi EG (ed) Hybrid Metaheuristics, Studies in Computational Intelligence, vol 434, Springer, pp 285–334

[5] Angel E, Bampis E (2005) A multi-start dynasearch algorithm for the time dependent single-machine total weighted tardiness scheduling problem. European Journal of Operational Research 162(1):281–289

[6] Aoun B, Boutaba R, Iraqi Y, Kenward G (2006) Gateway placement optimization in wireless mesh networks with QoS constraints. IEEE Journal on Selected Areas in Communications 24(11):2127–2136

[7] Applegate D, Cook W (1991) A computational study of the job-shop scheduling problem. ORSA Journal on Computing 3(2):149–156

[8] Applegate D, Bixby R, Chvátal V, Cook W (1999) Finding tours in the TSP. Tech. rep., Forschungsinstitut für Diskrete Mathematik, University of Bonn, Germany

[9] Archetti C, Guastaroba G, Speranza MG (2014) An ILP-refined tabu search for the directed profitable rural postman problem. Discrete Applied Mathematics 163, Part 1:3–16

[10] Bäck T (1996) Evolutionary Algorithms in Theory and Practice. Oxford University Press, New York

[11] Bäck T, Fogel D, Michalewicz Z (1997) Handbook of Evolutionary Computation. Oxford University Press, New York, NY

[12] Battiti R, Protasi M (1997) Reactive search, a history-sensitive heuristic for MAX-SAT. ACM Journal of Experimental Algorithmics 2:Article 2

[13] Battiti R, Tecchiolli G (1994) The reactive tabu search. ORSA Journal on Computing 6:126–140

[14] Bean JC (1994) Genetic algorithms and random keys for sequencing and optimization. ORSA Journal on Computing 6:154–160

[15] Benders JF (1962) Partitioning procedures for solving mixed-variables programming problems. Numerische Mathematik 4:238–252

[16] Bertsimas D, Demir R (2002) An approximate dynamic programming approach to multidimensional knapsack problems. Management Science 48(4):550–565

[17] Bertsimas D, Tsitsiklis JN (1997) Introduction to Linear Optimization. Athena Scientific

[18] Biesinger B, Schauer C, Hu B, Raidl GR (2013) Enhancing a genetic algorithm with a solution archive to reconstruct cross cut shredded text docu-

ments. In: Moreno-Díaz R, Pichler F, Quesada-Arencibia A (eds) Computer Aided Systems Theory – EUROCAST 2013, Springer, LNCS, vol 8111, pp 380–387

[19] Biesinger B, Hu B, Raidl GR (2014) An evolutionary algorithm for the leader-follower facility location problem with proportional customer behavior. In: Conference Proceedings of Learning and Intelligent Optimization Conference (LION 8), Springer, LNCS, vol 8426, pp 203–217

[20] Biesinger B, Hu B, Raidl G (2015) A hybrid genetic algorithm with solution archive for the discrete $(r|p)$-centroid problem. Journal of Heuristics 21(3):391–431

[21] Biesinger B, Hu B, Raidl G (2015) Models and algorithms for competitive facility location problems with different customer behavior. Annals of Mathematics and Artificial Intelligence DOI 10.1007/s10472-014-9448-0

[22] Biesinger B, Hu B, Raidl GR (2015) A variable neighborhood search for the generalized vehicle routing problem with stochastic demands. In: Ochoa G, Chicano F (eds) Evolutionary Computation in Combinatorial Optimization – EvoCOP 2015, Springer, LNCS, vol 9026, pp 48–60

[23] Blesa Aguilera MJ, Blum C, Cotta C, Fernández AJ, Gallardo JE, Roli A, Sampels M (eds) (2008) Proceedings of HM 2008 – Fifth International Workshop on Hybrid Metaheuristics, LNCS, vol 5296. Springer

[24] Blesa Aguilera MJ, Blum C, Di Gaspero L, Roli A, Sampels M, Schaerf A (eds) (2009) Proceedings of HM 2009 – Sixth International Workshop on Hybrid Metaheuristics, LNCS, vol 5818. Springer

[25] Blum C (2005) Beam-ACO: Hybridizing ant colony optimization with beam search: An application to open shop scheduling. Computers & Operations Research 32(6):1565–1591

[26] Blum C (2008) Beam-ACO for simple assembly line balancing. INFORMS Journal on Computing 20(4):618–627

[27] Blum C (2010) Beam-ACO for the longest common subsequence problem. In: Fogel G, et al. (eds) Proceedings of CEC 2010 – Congress on Evolutionary Computation, IEEE Press, Piscataway, NJ, vol 2, pp 1–8

[28] Blum C, Blesa MJ (2009) Solving the KCT problem: Large-scale neighborhood search and solution merging. In: Alba E, Blum C, Isasi P, León C, Gómez JA (eds) Optimization Techniques for Solving Complex Problems, Wiley & Sons, pp 407–421

[29] Blum C, Dorigo M (2004) The hyper-cube framework for ant colony optimization. IEEE Transactions on Man, Systems and Cybernetics – Part B 34(2):1161–1172

[30] Blum C, Mastrolilli M (2007) Using branch & bound concepts in construction-based metaheuristics: Exploiting the dual problem knowledge. In: Bartz-Beielstein T, Blesa Aguilera MJ, Blum C, Naujoks B, Roli A, Rudolph G, Sampels M (eds) Proceedings of HM 2007 – Fourth International Workshop on Hybrid Metaheuristics, Springer, LNCS, vol 4771, pp 123–139

[31] Blum C, Roli A (2003) Metaheuristics in combinatorial optimization: Overview and conceptual comparison. ACM Computing Surveys 35(3):268–308

[32] Blum C, Cotta C, Fernández AJ, Gallardo JE (2007) A probabilistic beam search algorithm for the shortest common supersequence problem. In: Cotta C, van Hemert JI (eds) Proceedings of EvoCOP 2007 – Seventh European Conference on Evolutionary Computation in Combinatorial Optimization, Springer, LNCS, vol 4446, pp 36–47

[33] Blum C, Bautista J, Pereira J (2008) An extended Beam-ACO approach to the time and space constrained simple assembly line balancing problem. In: van Hemert JI, Cotta C (eds) Proceedings of EvoCOP 2008 – Eighth European Conference on Evolutionary Computation in Combinatorial Optimization, Springer, LNCS, vol 4972, pp 85–96

[34] Blum C, Blesa Aguilera MJ, Roli A, Sampels M (eds) (2008) Hybrid Metaheuristics – An Emerging Approach to Optimization, Studies in Computational Intelligence, vol 114. Springer

[35] Blum C, Blesa MJ, Calvo B (2014) Beam-ACO for the repetition-free longest common subsequence problem. In: Legrand P, Corsini MM, Hao JK, Monmarché N, Lutton E, Schoenauer M (eds) Proceedings of EA 2013 – 11th Conference on Artificial Evolution, Springer, LNCS, vol 8752, pp 79–90

[36] Blum C, Lozano JA, Pinacho Davidson P (2014) Iterative probabilistic tree search for the minimum common string partition problem. In: Blesa MJ, Blum C, Voss S (eds) Proceedings of HM 20104– 9th International Workshop on Hybrid Metaheuristics, Springer, LNCS, vol 8457, pp 154–154

[37] Blum C, Lozano JA, Pinacho Davidson P (2015) Mathematical programming strategies for solving the minimum common string partition problem. European Journal of Operational Research 242(3):769–777

[38] Boettcher S, Percus AG (2001) Extremal optimization for graph partitioning. Physical Review E 64(2):026,114

[39] Borisovsky P, Dolgui A, Eremeev A (2009) Genetic algorithms for a supply management problem: MIP-recombination vs. greedy decoder. European Journal of Operational Research 195(3):770–779

[40] Boschetti M, Maniezzo V (2009) Benders decomposition, Lagrangian relaxation and metaheuristic design. Journal of Heuristics 15:283–312

[41] Boschetti M, Maniezzo V, Roffilli M (2009) Decomposition techniques as metaheuristic frameworks. In: Maniezzo V, Stützle T, Voss S (eds) Matheuristics – Hybridizing Metaheuristics and Mathematical Programming, Annals of Information Systems, vol 10, Springer, pp 135–158

[42] Bouamama S, Blum C, Boukerram A (2012) A population-based iterated greedy algorithm for the minimum weight vertex cover problem. Applied Soft Computing 12(6):1632–1639

[43] Boyer V, Elkihel M, El Baz D (2009) Heuristics for the 0–1 multidimensional knapsack problem. European Journal of Operational Research 199(3):658–664

[44] Burke EK, Kendall G, Newall J, Hart E, Ross P, Schulenburg S (2003) Hyper-heuristics: An emerging direction in modern search technology. In: Glover F, Kochenberger G (eds) Handbook of Metaheuristics, International Series

in Operations Research & Management Science, vol 57, Kluwer Academic Publishers, pp 457–474

[45] Burke EK, Gendreau M, Hyde M, Kendall G, Ochoa G, Özcan E, Qu R (2013) Hyper-heuristics: A survey of the state of the art. Journal of the Operational Research Society pp 1695–1724

[46] Cahon S, Melab N, Talbi EG (2004) ParadisEO: A framework for the reusable design of parallel and distributed metaheuristics. Journal of Heuristics 10(3):357–380

[47] Caldeira J, Azevedo R, Silva CA, da Costa Sousa JM (2007) Beam-ACO distributed optimization applied to supply-chain management. In: Melin P, Castillo O, Aguilar LT, Kacprzyk J, Pedrycz W (eds) Proceedings of IFSA 2007 – 12th International Fuzzy Systems Association World Congress on Foundations of Fuzzy Logic and Soft Computing, Springer, LNCS, vol 4529, pp 799–809

[48] Caldeira J, Azevedo R, Silva CA, da Costa Sousa JM (2007) Supply-chain management using ACO and Beam-ACO algorithms. In: Proceedings of FUZZ-IEEE 2007 – IEEE International Conference on Fuzzy Systems, IEEE press, pp 1–6

[49] Calégary P, Coray G, Hertz A, Kobler D, Kuonen P (1999) A taxonomy of evolutionary algorithms in combinatorial optimization. Journal of Heuristics 5:145–158

[50] Cambazard H, Hebrard E, O'Sullivan B, Papadopoulos A (2012) Local search and constraint programming for the post enrolment-based course timetabling problem. Annals of Operations Research 194(1):111–135

[51] Caserta M, Voß S, Sniedovich M (2011) Applying the corridor method to a blocks relocation problem. OR Spectrum 33(4):915–929

[52] Černý V (1985) A thermodynamical approach to the travelling salesman problem: An efficient simulation algorithm. Journal of Optimization Theory and Applications 45:41–51

[53] Chen WN, Zhang J, Chung HSH, Zhong WL, Wu WG, Shi YH (2010) A novel set-based particle swarm optimization method for discrete optimization problems. IEEE Transactions on Evolutionary Computation 14(2):278–300

[54] Chen X, Zheng J, Fu Z, Nan P, Zhong Y, Lonardi S, Jiang T (2005) Computing the assignment of orthologous genes via genome rearrangement. In: Proceedings of the Asia Pacific Bioinformatics Conference 2005, pp 363–378

[55] Chen YP, Liestman AL (2002) Approximating minimum size weakly-connected dominating sets for clustering mobile ad hoc networks. In: Proceedings of MobiHoc 2002 – 3rd ACM International Symposium on Mobile Ad Hoc Networking & Computing, ACM, New York, NY, pp 165–172

[56] Chrobak M, Kolman P, Sgall J (2004) The greedy algorithm for the minimum common string partition problem. In: Jansen K, Khanna S, Rolim JDP, Ron D (eds) Proceedings of APPROX 2004 – 7th International Workshop on Approximation Algorithms for Combinatorial Optimization Problems, LNCS, vol 3122, Springer, pp 84–95

[57] Chu PC, Beasley JE (1998) A genetic algorithm for the multidimensional knapsack problem. Journal of Heuristics 4:63–86

[58] Clark BN, Colbourn CJ, Johnson DS (1990) Unit disk graphs. Discrete Mathematics 86(1-3)

[59] Clements D, Crawford J, Joslin D, Nemhauser G, Puttlitz M, Savelsbergh M (1997) Heuristic optimization: A hybrid AI/OR approach. In: Davenport A, Beck C (eds) Proceedings of the Workshop on Industrial Constraint-Directed Scheduling, held in conjunction with the Third International Conference on Principles and Practice of Constraint Programming (CP97)

[60] Clerc M (ed) (2006) Particle Swarm Optimization. Wiley-ISTE Publishers, Newport Beach, CA

[61] Codato G, Fischetti M (2006) Combinatorial Benders' cuts for mixed integer linear programming. Operations Research 54(4):756–766

[62] Congram RK, Potts CN, van de Velde SL (2002) An iterated dynasearch algorithm for the single-machine total weighted tardiness scheduling problem. INFORMS Journal on Computing 14(1):52–67

[63] Cook W, Seymour P (2003) Tour merging via branch-decomposition. INFORMS Journal on Computing 15(3):233–248

[64] Cordeau JF, Stojković G, Soumis F, Desrosiers J (2001) Benders decomposition for simultaneous aircraft routing and crew scheduling. Transportation Science 35(4):375–388

[65] Cormen TH, Leiserson CE, Rivest RL, Stein C (2009) Introduction to Algorithms, 3rd edn. MIT Press

[66] Cormode G, Muthukrishnan S (2007) The string edit distance matching problem with moves. ACM Transactions on Algorithms 3(2):1–19

[67] Cotta C (1998) A study of hybridisation techniques and their application to the design of evolutionary algorithms. AI Communications 11(3–4):223–224

[68] Cotta C, Troya JM (2003) Embedding branch and bound within evolutionary algorithms. Applied Intelligence 18:137–153

[69] Cotta C, Talbi EG, Alba E (2005) Parallel hybrid metaheuristics. In: Alba E (ed) Parallel Metaheuristics: A New Class of Algorithms, Wiley, pp 347–370

[70] Coudert D, Nepomuceno N, Rivano H (2010) Power-efficient radio configuration in fixed broadband wireless networks. Computer Communications 33(8):898–906

[71] Coudert D, Nepomuceno NV, Tahiri I (2011) Energy saving in fixed wireless broadband networks. In: Pahl J, Reiners T, Voß S (eds) Proceedings of INOC 2011 – 5th International Conference on Network Optimization, Lecture Notes in Computer Science, vol 6701, Springer, pp 484–489

[72] Danna E, Rothberg E, Le Pape C (2005) Exploring relaxation induced neighborhoods to improve MIP solutions. Mathematical Programming, Series A 102:71–90

[73] Dantzig GB, Wolfe P (1960) Decomposition principle for linear programs. Operations Research 8:101–111

[74] De Franceschi R, Fischetti M, Toth P (2006) A new ILP-based refinement heuristic for vehicle routing problems. Mathematical Programming, Series B 105(2):471–499
[75] Deb K (2008) Multi-Objective Optimization Using Evolutionary Algorithms. Wiley
[76] Desaulniers G, Desrosiers J, Solomon MM (2005) Column Generation. Springer
[77] Dorigo M, Gambardella LM (1997) Ant Colony System: A cooperative learning approach to the traveling salesman problem. IEEE Transactions on Evolutionary Computation 1(1):53–66
[78] Dorigo M, Stützle T (2004) Ant Colony Optimization. MIT Press, Cambridge, MA
[79] Dorigo M, Stützle T (2010) Ant colony optimization: Overview and recent advances. In: Gendreau M, Potvin JY (eds) Handbook of Metaheuristics, International Series in Operations Research & Management Science, vol 146, 2nd edn, Springer, pp 227–264
[80] Dörner K, et al. (eds) (2010) Proceedings of Matheuristics 2010: Third International Workshop on Model Based Metaheuristics, Vienna, Austria
[81] Dror M, Haouari M, Chaouachi JS (2000) Generalized spanning trees. European Journal of Operational Research 120:583–592
[82] Dueck G (1993) New optimization heuristics: the great deluge algorithm and the record-to-record travel. Journal of Computational Physics 104(1):86–92
[83] Dumitrescu I, Stützle T (2003) Combinations of local search and exact algorithms. In: Cagnoni S, Johnson CG, Romero Cardalda JJ, Marchiori E, Corne DW, Meyer JA, Gottlieb J, Middendorf M, Guillot A, Raidl GR, Hart E (eds) Applications of Evolutionary Computation, Springer, LNCS, vol 2611, pp 211–223
[84] Easton T, Singireddy A (2008) A large neighborhood search heuristic for the longest common subsequence problem. Journal of Heuristics 14(3):271–283
[85] Easwaran G, Üster H (2009) Tabu search and Benders decomposition approaches for a capacitated closed-loop supply chain network design problem. Transportation Science 43(3):301–320
[86] Ehrgott M, Gandibleux X (2008) Hybrid metaheuristics for multi-objective combinatorial optimization. In: Blum C, Blesa Aguilera MJ, Roli A, Sampels M (eds) Hybrid Metaheuristics – An Emerging Approach to Optimization, Springer, Studies in Computational Intelligence, vol 114, pp 221–259
[87] Engelbrecht AP (2005) Fundamentals of Computational Swarm Intelligence. Wiley & Sons
[88] Erel E, Sabuncuoglu I, Sekerci H (2005) Stochastic assembly line balancing using beam search. International Journal of Production Research 43(7):1411–1426
[89] Eremeev AV (2008) On complexity of optimal recombination for binary representations of solutions. Evolutionary Computation 16(1):127–147
[90] Eremeev AV, Kovalenko JV (2013) Optimal recombination in genetic algorithms. ArXiv technical report http://arxivorg/abs/13075519

[91] Feller W (1968) An Introduction to Probability Theory and its Applications, 3rd edn. John Wiley

[92] Feo TA, Resende MGC (1995) Greedy randomized adaptive search procedures. Journal of Global Optimization 6:109–133

[93] Ferdous SM, Sohel Rahman M (2013) Solving the minimum common string partition problem with the help of ants. In: Tan Y, Shi Y, Mo H (eds) Proceedings of ICSI 2013 – 4th International Conference on Advances in Swarm Intelligence, LNCS, vol 7928, Springer, pp 306–313

[94] Ferdous SM, Sohel Rahman M (2014) A MAX-MIN ant colony system for minimum common string partition problem. CoRR abs/1401.4539, http://arxiv.org/abs/1401.4539

[95] Feremans C (2001) Generalized spanning trees and extensions. PhD thesis, Université libre de Bruxelles, Brussels, Belgium

[96] Feremans C, Labbé M, Laporte G (2002) A comparative analysis of several formulations for the generalized minimum spanning tree problem. Networks 39(1):29–34

[97] Feremans C, Labbé M, Laporte G (2004) The generalized minimum spanning tree problem: Polyhedral analysis and branch-and-cut algorithm. Networks 43(2):71–86

[98] Fischetti M, Lodi A (2003) Local branching. Mathematical Programming, Series B 98:23–47

[99] Fisher ML (1981) The Lagrangian relaxation method for solving integer programming problems. Management Science 27(1):1–18

[100] Flindt Muller L, Spoorendonk S, Pisinger D (2012) A hybrid adaptive large neighborhood search heuristic for lot-sizing with setup times. European Journal of Operational Research 218(3):614–623

[101] Fogel DB (1994) An introduction to simulated evolutionary optimization. IEEE Transactions on Neural Networks 5(1):3–14

[102] Fogel LJ (1962) Toward inductive inference automata. In: Proceedings of the International Federation for Information Processing Congress, pp 395–399

[103] Fogel LJ, Owens AJ, Walsh MJ (1966) Artificial Intelligence through Simulated Evolution. Wiley

[104] Freville A, Plateau G (1994) An efficient preprocessing procedure for the multidimensional 0–1 knapsack problem. Discrete Applied Mathematics 49(1):189–212

[105] Gallardo JE (2012) A multilevel probabilistic beam search algorithm for the shortest common supersequence problem. PLOS ONE 7(12)

[106] Gantovnik VB, Anderson-Cook CM, Gürdal Z, Watson LT (2003) A genetic algorithm with memory for mixed discrete-continuous design optimization. Computers and Structures 81:2003–2009

[107] Garey MR, Johnson DS (1979) Computers and Intractability: A Guide to the Theory of NP-Completeness. W. H. Freeman, New York

[108] Ghirardi M, Potts CN (2005) Makespan minimization for scheduling unrelated parallel machines: A recovering beam search approach. European Journal of Operational Research 165(2):457–467

[109] Ghosh D (2003) Solving medium to large sized Euclidean generalized minimum spanning tree problems. Tech. Rep. NEP-CMP-2003-09-28, Indian Institute of Management, Research and Publication Department, Ahmedabad, India

[110] Gilmour S, Dras M (2006) Kernelization as heuristic structure for the vertex cover problem. In: Dorigo M, Gambardella LM, Birattari M, Martinoli A, Poli R, Stützle T (eds) Proceedings of ANTS 2006 – 5th International Workshop on Ant Colony Optimization and Swarm Intelligence, Springer, LNCS, vol 4150, pp 452–459

[111] Glover F (1990) Tabu search – part II. ORSA Journal on Computing 2(1):4–32

[112] Glover F (2006) Parametric tabu-search for mixed integer programming. Computers and Operations Research 33(9):2449–2494

[113] Glover F, Kochenberger G (eds) (2003) Handbook of Metaheuristics, International Series in Operations Research & Management Science, vol 57. Kluwer Academic Publishers

[114] Glover F, Laguna M (1997) Tabu Search. Kluwer Academic Publishers

[115] Glover F, Laguna M, Martí R (2000) Fundamentals of scatter search and path relinking. Control and Cybernetics 39(3):653–684

[116] Goldberg DE (1989) Genetic Algorithms in Search, Optimization, and Learning. Addison-Wesley, Reading, MA

[117] Golden B, Raghavan S, Stanojevic D (2005) Heuristic search for the generalized minimum spanning tree problem. INFORMS Journal on Computing 17(3):290–304

[118] Goldstein A, Kolman P, Zheng J (2005) Minimum common string partition problem: Hardness and approximations. In: Fleischer R, Trippen G (eds) Proceedings of ISAAC 2004 – 15th International Symposium on Algorithms and Computation, LNCS, vol 3341, Springer, pp 484–495

[119] Goldstein I, Lewenstein M (2011) Quick greedy computation for minimum common string partitions. In: Giancarlo R, Manzini G (eds) Proceedings of CPM 2011 – 22nd Annual Symposium on Combinatorial Pattern Matching, LNCS, vol 6661, Springer, pp 273–284

[120] Gomes TM, Santos HG, Souza JF (2013) A pre-processing aware RINS based MIP heuristic. In: Blesa MJ, Blum C, Festa P, Roli A, Sampels M (eds) Proceedings of HM 2013 – Eighth International Workshop on Hybrid Metaheuristics, Springer, LNCS, vol 7919, pp 1–11

[121] Gonçalves JF, Resende MGC (2011) Biased random-key genetic algorithms for combinatorial optimization. Journal of Heuristics 17(5):487–525

[122] Grosso A, Della Croce F, Tadei R (2004) An enhanced dynasearch neighborhood for the single-machine total weighted tardiness scheduling problem. Operations Research Letters 32(1):68–72

[123] Gusfield D (1997) Algorithms on Strings, Trees, and Sequences. Cambridge University Press

[124] Han B, Jia W (2007) Clustering wireless ad hoc networks with weakly connected dominating set. Journal of Parallel and Distributed Computing 67(6):727–737
[125] Hanafi S, Freville A (1998) An efficient tabu search approach for the 0–1 multidimensional knapsack problem. European Journal of Operational Research 106(2–3):659–675
[126] Hansen P, Mladenović N (1999) An introduction to variable neighborhood search. In: Voß S, Martello S, Osman I, Roucairol C (eds) Meta-heuristics: advances and trends in local search paradigms for optimization, Kluwer Academic Publishers, pp 433–438
[127] Hansen P, Mladenović N (2001) Variable neighborhood search: Principles and applications. European Journal of Operational Research 130:449–467
[128] Hansen P, Mladenović N, Urosević D (2006) Variable neighborhood search and local branching. Computers & Operations Research 33(10):3034–3045
[129] Hansen P, Maniezzo V, Fischetti M, Stützle T (eds) (2008) Proceedings of Matheuristics 2008: Second International Workshop on Model Based Metaheuristics, Bertinoro, Italy
[130] Haouari M, Siala JC (2006) A hybrid Lagrangian genetic algorithm for the prize collecting Steiner tree problem. Computers & Operations Research 33(5):1274–1288
[131] He D (2007) A novel greedy algorithm for the minimum common string partition problem. In: Mandoiu I, Zelikovsky A (eds) Proceedings of ISBRA 2007 – Third International Symposium on Bioinformatics Research and Applications, LNCS, vol 4463, Springer, pp 441–452
[132] Hertz A, Kobler D (2000) A framework for the description of evolutionary algorithms. European Journal of Operational Research 126:1–12
[133] Holland JH (1975) Adaptation in Natural and Artificial Systems. The University of Michigan Press, Ann Arbor, MI
[134] Hooker J (2012) Integrated Methods for Optimization, Int. Series in Operations Research & Management Science, vol 170, 2nd edn. Springer
[135] Hooker JN (2007) Planning and scheduling by logic-based Benders decomposition. Operations Research 55(3):588–602
[136] Hooker JN, Ottosson G (2003) Logic-based Benders decomposition. Mathematical Programming 96:33–60
[137] Hoos H, Stützle T (2005) Stochastic Local Search – Foundations and Applications. Morgan Kaufmann Publishers
[138] Hu B (2008) Hybrid metaheuristics for generalized network design problems. PhD thesis, Vienna University of Technology, Institute of Computer Graphics and Algorithms, Vienna, Austria
[139] Hu B, Raidl GR (2008) Effective neighborhood structures for the generalized traveling salesman problem. In: van Hemert JI, Cotta C (eds) Evolutionary Computation in Combinatorial Optimization – EvoCOP 2008, Springer, LNCS, vol 4972, pp 36–47
[140] Hu B, Raidl GR (2012) An evolutionary algorithm with solution archive for the generalized minimum spanning tree problem. In: Moreno-Díaz R, Pichler

F, Quesada-Arencibia A (eds) Proceedings of the 13th International Conference on Computer Aided Systems Theory: Part I, Springer, LNCS, vol 6927, pp 287–294

[141] Hu B, Raidl GR (2012) An evolutionary algorithm with solution archives and bounding extension for the generalized minimum spanning tree problem. In: Proceedings of the 14th Annual Conference on Genetic and Evolutionary Computation (GECCO), ACM Press, Philadelphia, PA, USA, pp 393–400

[142] Hu B, Leitner M, Raidl GR (2008) Combining variable neighborhood search with integer linear programming for the generalized minimum spanning tree problem. Journal of Heuristics 14(5):473–499

[143] Ihler E, Reich G, Widmayer P (1999) Class Steiner trees and VLSI-design. Discrete Applied Mathematics 90:173–194

[144] Jaśkowski W, Szubert M, Gawron P (2015) A hybrid MIP-based large neighborhood search heuristic for solving the machine reassignment problem. Annals of Operations Research DOI 10.1007/s10479-014-1780-6

[145] Jeet V, Kutanoglu E (2007) Lagrangian relaxation guided problem space search heuristic for generalized assignment problems. European Journal of Operational Research 182(3):1039–1056

[146] Jiang H, Chen Y (2010) An efficient algorithm for generalized minimum spanning tree problem. In: GECCO 2010: Proceedings of the 12th Annual Conference on Genetic and Evolutionary Computation, ACM, pp 217–224

[147] Joslin DE, Clements DP (1999) "Squeaky Wheel" optimization. Journal of Artificial Intelligence Research 10:353–373

[148] Jovanovic R, Tuba M, Simian D (2010) Ant colony optimization applied to minimum weight dominating set problem. In: Proceedings of the 12th WSEAS International Conference on Automatic Control, Modelling & Simulation, pp 322–326

[149] Kann V (1992) On the approximability of NP-complete optimization problems. PhD thesis, Royal Institute of Technology, Stockholm, Sweden

[150] Kaplan H, Shafrir N (2006) The greedy algorithm for edit distance with moves. Information Processing Letters 97(1):23–27

[151] Karaboga D, Basturk B (2007) A powerful and efficient algorithm for numerical function optimization: artificial bee colony (ABC) algorithm. Journal of Global Optimization 39(3):459–471

[152] Karaboga D, Basturk B (2008) On the performance of artificial bee colony (ABC) algorithm. Applied Soft Computing 8(1):687–697

[153] Kauffman SA (1993) The Origins of Order: Self-Organization and Selection in Evolution. Oxford University Press

[154] Kellerer H, Pferschy U, Pisinger D (2004) Knapsack Problems. Springer

[155] Kennedy J, Eberhart RC (1995) Particle swarm optimization. In: Proceedings of the 1995 IEEE International Conference on Neural Networks, IEEE Press, Piscataway, NJ, vol 4, pp 1942–1948

[156] Kennedy J, Eberhart RC, Shi Y (2004) Swarm Intelligence. Morgan Kaufmann Publishers, San Francisco, CA

[157] Kirkpatrick S, Gellat C, Vecchi M (1983) Optimization by simulated annealing. Science 220:671–680

[158] Klau GW, Ljubić I, Moser A, Mutzel P, Neuner P, Pferschy U, Raidl G, Weiskircher R (2004) Combining a memetic algorithm with integer programming to solve the prize-collecting Steiner tree problem. In: Proceedings of GECCO 2004 – Genetic and Evolutionary Computation Conference, Springer, LNCS, vol 3102, pp 1304–1315

[159] Kleinberg J, Tardos É (2005) Algorithm Design. Addison-Wesley

[160] Knuth DE (1973) The Art of Computer Programming Vol. III: Sorting and Searching. Addison-Wesley

[161] Kolman P (2005) Approximating reversal distance for strings with bounded number of duplicates. In: Jedrzejowicz J, Szepietowski A (eds) Proceedings of MFCS 2005 – 30th International Symposium on Mathematical Foundations of Computer Science, LNCS, vol 3618, Springer, pp 580–590

[162] Kolman P, Waleń T (2007) Reversal distance for strings with duplicates: Linear time approximation using hitting set. In: Erlebach T, Kaklamanis C (eds) Proceedings of WAOA 2007 – 4th International Workshop on Approximation and Online Algorithms, LNCS, vol 4368, Springer, pp 279–289

[163] Kong M, Tian P, Kao Y (2008) A new ant colony optimization algorithm for the multidimensional knapsack problem. Computers & Operations Research 35(8):2672–2683

[164] Kong X, Gao L, Ouyang H, Li S (2015) Solving large-scale multidimensional knapsack problems with a new binary harmony search algorithm. Computers & Operations Research 63:7–22

[165] Koza JR (1992) Genetic Programming: On the Programming of Computers by Means of Natural Selection. MIT Press, Cambridge, MA

[166] Kratica J (1999) Improving performances of the genetic algorithm by caching. Computers and Artificial Intelligence 18(3):271–283

[167] Kruskal JB (1956) On the shortest spanning subtree and the traveling salesman problem. In: Proceedings of the American Mathematical Society, vol 7, pp 48–50

[168] Lai MC, Sohn HS, Tseng TL, Chiang C (2010) A hybrid algorithm for capacitated plant location problem. Expert Systems with Applications 37(12):8599–8605

[169] Lai MC, Sohn HS, Tseng TL, Bricker DL (2012) A hybrid Benders/genetic algorithm for vehicle routing and scheduling problem. International Journal of Industrial Engineering 19(1):33–46

[170] Lazić J, Hanafi S, Mladenović N, Urošević D (2010) Variable neighbourhood decomposition search for 0–1 mixed integer programs. Computers & Operations Research 37(6):1055–1067

[171] Leitner M, Raidl GR (2008) Lagrangian decomposition, metaheuristics, and hybrid approaches for the design of the last mile in fiber optic networks. In: Blesa Aguilera MJ, Blum C, Cotta C, Fernández AJ, Gallardo JE, Roli A, Sampels M (eds) Proceedings of HM 2008 – Fifth International Workshop on Hybrid Metaheuristics, Springer, LNCS, vol 5296, pp 158–174

[172] Leitner M, Hu B, Raidl GR (2007) Variable neighborhood search for the generalized minimum edge biconnected network problem. In: Fortz B (ed) Proceedings of the International Network Optimization Conference 2007, Spa, Belgium, paper number 69
[173] Leung SCH, Zhang D, Zhou C, Wu T (2012) A hybrid simulated annealing metaheuristic algorithm for the two-dimensional knapsack packing problem. Computers & Operations Research 39(1):64–73
[174] Li VC, Liang YC, Chang HF (2012) Solving the multidimensional knapsack problems with generalized upper bound constraints by the adaptive memory projection method. Computers & Operations Research 39(9):2111–2121
[175] Lodi A, Milano M, Toth P (eds) (2010) Proceedings of CPAIOR 2010 – 7th International Conference on the Integration of AI and OR Techniques in Constraint Programming for Combinatorial Optimization Problems, LNCS, vol 6140. Springer
[176] López-Ibáñez M, Blum C (2009) Beam-ACO based on stochastic sampling: A case study on the TSP with time windows. In: Stützle T (ed) Proceedings of LION 3 – 3rd International Conference on Learning and Intelligent Optimization, Springer, LNCS, vol 5851, pp 59–73
[177] López-Ibáñez M, Blum C, Thiruvady DR, Ernst AT, Meyer B (2009) Beam-ACO based on stochastic sampling for makespan optimization concerning the TSP with time windows. In: Cotta C, Cowling PI (eds) Proceedings of EvoCOP 2009 – 9th European Conference on Evolutionary Computation in Combinatorial Optimization, Springer, LNCS, vol 5482, pp 97–108
[178] López-Ibáñez M, Blum C (2010) Beam-ACO for the travelling salesman problem with time windows. Computers & Operations Research 37(9):1570–1583
[179] López-Ibáñez M, Dubois-Lacoste J, Stützle T, Birattari M (2011) The irace package, iterated race for automatic algorithm configuration. Tech. Rep. TR/IRIDIA/2011-004, IRIDIA, Université libre de Bruxelles, Belgium
[180] Louis SJ, Li G (1997) Combining robot control strategies using genetic algorithms with memory. In: Evolutionary Programming VI, Springer, LNCS, vol 1213, pp 431–442
[181] Lourenço HR, Martin O, Stützle T (2002) Iterated local search. In: Glover F, Kochenberger G (eds) Handbook of Metaheuristics, International Series in Operations Research & Management Science, vol 57, Kluwer Academic Publishers, Norwell, MA, pp 321–353
[182] Lozano M, García-Martínez C (2010) Hybrid metaheuristics with evolutionary algorithms specializing in intensification and diversification: Overview and progress report. Computers & Operations Research 37(3):481–497
[183] Lübbecke ME, Desrosiers J (2005) Selected topics in column generation. Operations Research 53(6):1007–1023
[184] Maniezzo V (1999) Exact and approximate nondeterministic tree-search procedures for the quadratic assignment problem. INFORMS Journal on Computing 11(4):358–369

[185] Maniezzo V, Carbonaro A (2000) An ANTS heuristic for the frequency assignment problem. Future Generation Computer Systems 16:927–935

[186] Maniezzo V, Milandri M (2002) An ant-based framework for very strongly constrained problems. In: Dorigo M, Di Caro G, Sampels M (eds) Proceedings of ANTS 2002 – 3rd International Workshop on Ant Algorithms, Springer, LNCS, vol 2463, pp 222–227

[187] Maniezzo V, Hansen P, Voss S (eds) (2006) Proceedings of Matheuristics 2006: First International Workshop on Mathematical Contributions to Metaheuristics, Bertinoro, Italy

[188] Maniezzo V, Stützle T, Voss S (eds) (2009) Matheuristics – Hybridizing Metaheuristics and Mathematical Programming, Annals of Information Systems, vol 10. Springer

[189] Marino A, Prügel-Bennett A, Glass CA (1999) Improving graph colouring with linear programming and genetic algorithms. In: Miettinen K, Makela MM, Toivanen J (eds) Proceedings of EUROGEN 99 – the Int. Conference on Evolutionary and Deterministic Methods for Design, Optimization and Control with Applications to Industrial and Societal Problems, Jyväskylä, Finland, pp 113–118

[190] Massen F, Deville Y, Hentenryck PV (2012) Pheromone-based heuristic column generation for vehicle routing problems with black box feasibility. In: Beldiceanu N, Jussien N, Pinson É (eds) Integration of AI and OR Techniques in Constraint Programming for Combinatorial Optimization Problems – CPAIOR 2012, Springer, LNCS, vol 7298, pp 260–274

[191] Massen F, López-Ibáñez M, Stützle T, Deville Y (2013) Experimental analysis of pheromone-based heuristic column generation using irace. In: Blesa MJ, Blum C, Festa P, Roli A, Sampels M (eds) Proceedings of HM 2013 – Eighth International Workshop on Hybrid Metaheuristics, Springer, LNCS, vol 7919, pp 92–106

[192] Mastrolilli M, Blum C (2010) On the use of different types of knowledge in metaheuristics based on constructing solutions. Engineering Applications of Artificial Intelligence 23(5):650–659

[193] Mauldin ML (1984) Maintaining diversity in genetic search. In: Proceedings of the 4th National Conference on Artificial Intelligence, Austin, Texas, pp 247–250

[194] Metropolis N, Rosenbluth A, Rosenbluth M, Teller A, Teller E (1953) Equation of state calculations by fast computing machines. Journal of Chemical Physics 21:1087–1092

[195] Michalewicz Z, Michalewicz M (1997) Evolutionary computation techniques and their applications. In: Proceedings of the IEEE International Conference on Intelligent Processing Systems, IEEE, Beijing, China, pp 14–24

[196] Michalewicz Z, Siarry P (2008) Special issue on adaptation of discrete metaheuristics to continuous optimization. European Journal of Operational Research 185:1060–1273

[197] Mitchell M (1998) An Introduction to Genetic Algorithms. MIT Press, Cambridge, MA

[198] Mitchell M, Forrest S, Holland JH (1992) The Royal Road for genetic algorithms: Fitness landscapes and GA performance. In: Varela FJ, Bourgine P (eds) Towards a Practice of Autonomous Systems: Proceedings of the First European Conference on Artificial Life, MIT Press, pp 245–254
[199] Mitrović-Minić S, Punnen AP (2010) Variable intensity local search. In: Maniezzo V, Stützle T, Voß S (eds) Matheuristics, Annals of Information Systems, vol 10, Springer, pp 245–252
[200] Molina D, Lozano M, García-Martínez C, Herrera F (2010) Memetic algorithms for continuous optimisation based on local search chains. Evolutionary Computation 18(1):27–63
[201] Moscato P (1999) Memetic algorithms: A short introduction. In: Corne D, Dorigo M, Glover F, Dasgupta D, Moscato P, Poli R, Price KV (eds) New Ideas in Optimization, McGraw-Hill, pp 219–234
[202] Myung YS, Lee CH, Tcha DW (1995) On the generalized minimum spanning tree problem. Networks 26:231–241
[203] Nemhauser GL, Wolsey LA (1988) Integer and Combinatorial Optimization. Wiley & Sons
[204] Nepomuceno N, Pinheiro P, Coelho ALV (2008) A hybrid optimization framework for cutting and packing problems. In: Cotta C, van Hemert J (eds) Recent Advances in Evolutionary Computation for Combinatorial Optimization, Studies in Computational Intelligence, vol 153, Springer, pp 87–99
[205] Nepomuceno NV (2006) Combinação de metaheurísticas e programação linear inteira: uma metodologia híbrida aplicada ao problema de carregamento de contêineres. Master's thesis, University of Fortaleza, Brazil
[206] Nepomuceno NV, Pinheiro PR, Coelho ALV (2007) Combining metaheuristics and integer linear programming: A hybrid methodology applied to the container loading problem. In: Proceedings of the XX Congreso da Sociedade Brasileira de Computação, Concurso de Teses e Dissertações, pp 2028–2032
[207] Nepomuceno NV, Pinheiro PR, Coelho ALV (2007) Tackling the container loading problem: A hybrid approach based on integer linear programming and genetic algorithms. In: Cotta C, van Hemert J (eds) Proceedings of EvoCOP 2007 – 7th European Conference on Evolutionary Computation in Combinatorial Optimization, Lecture Notes in Computer Science, vol 4446, Springer, pp 154–165
[208] Neto T, Pedroso JP (2003) GRASP for linear integer programming. In: Sousa JP, Resende MGC (eds) Metaheuristics: Computer Decision Making, Combinatorial Optimization Book Series, Kluwer Academic Publishers, pp 545–574
[209] Öncan T, Cordeau JF, Laporte G (2008) A tabu search heuristic for the generalized minimum spanning tree problem. European Journal of Operational Research 191(2):306–319
[210] Öncan T, Kabadi SN, Nair KPN, Punnen AP (2008) VLSN search algorithms for partitioning problems using matching neighbourhoods. The Journal of the Operational Research Society 59:388–398

[211] Ow PS, Morton TE (1988) Filtered beam search in scheduling. International Journal of Production Research 26:297–307

[212] Palacios JJ, Vela CR, González-Rodríguez I, Puente J (2014) Genetic beam search for fuzzy open shop problems. In: Proceedings of META 2014 – 5th International Conference on Metaheuristics and Nature-Inspired Computing, Singapore

[213] Palmer CC, Kershenbaum A (1994) Representing trees in genetic algorithms. In: Schaffer D, et al. (eds) Proceedings of the First IEEE Conference on Evolutionary Computation, IEEE Press, pp 379–384

[214] Pan QK, Tasgetiren MF, Liang YC (2008) A discrete particle swarm optimization algorithm for the no-wait flowshop scheduling problem. Computers & Operations Research 35(9):2807–2839

[215] Pan QK, Tasgetiren MF, Suganthan PN, Chua TJ (2011) A discrete artificial bee colony algorithm for the lot-streaming flow shop scheduling problem. Information Sciences 181(12):2455–2468

[216] Papadimitriou CH, Steiglitz K (1982) Combinatorial Optimization - Algorithms and Complexity. Dover Publications, New York

[217] Pedroso JP (2005) Tabu search for mixed integer programming. In: Rego C, Alidaee B (eds) Metaheuristic Optimization via Memory and Evolution, Operations Research/Computer Science Interfaces Series, vol 30, Springer, pp 247–261

[218] Pereira MA, Coelho LC, Lorena LAN, de Souza LC (2015) A hybrid method for the probabilistic maximal covering location–allocation problem. Computers & Operations Research 57:51–59

[219] Perron L, Trick MA (eds) (2008) Proceedings of CPAIOR 2008 – 5th International Conference on the Integration of AI and OR Techniques in Constraint Programming for Combinatorial Optimization Problems, LNCS, vol 5015. Springer

[220] Perron L, Shaw P, Furnon V (2004) Propagation guided large neighborhood search. In: Wallace M (ed) Principles and Practice of Constraint Programming – CP 2004, Springer, LNCS, vol 3258, pp 468–481

[221] Pesant G, Gendreau M (1996) A view of local search in constraint programming. In: Freuder E (ed) Principles and Practice of Constraint Programming - CP'96, Springer, LNCS, vol 1118, pp 353–366

[222] Pesant G, Gendreau M (1999) A constraint programming framework for local search methods. Journal of Heuristics 5:255–279

[223] Pessoa LS, Resende MGC, Ribeiro CC (2013) A hybrid Lagrangean heuristic with GRASP and path-relinking for set *k*-covering. Computers & Operations Research 40(12):3132–3146

[224] Pinheiro PR, Coelho ALV, de Aguiar AB, Bonates TO (2011) On the concept of density control and its application to a hybrid optimization framework: Investigation into cutting problems. Computers & Industrial Engineering 61(3):463–472

[225] Pinheiro PR, Coelho ALV, de Aguiar AB, de Menezes Sobreira Neto A (2012) Towards aid by generate and solve methodology: application in the

problem of coverage and connectivity in wireless sensor networks. International Journal of Distributed Sensor Networks 2012, article ID 790459

[226] Pirkwieser S, Raidl GR (2010) Matheuristics for the periodic vehicle routing problem with time windows. In: Proceedings of Matheuristics 2010: Third International Workshop on Model-Based Metaheuristics, Vienna, Austria, pp 83–95

[227] Pirkwieser S, Raidl GR (2010) Variable neighborhood search coupled with ILP-based large neighborhood searches for the (periodic) location-routing problem. In: Blesa Aguilera MJ, Blum C, Raidl GR, Roli A, Sampels M (eds) Proceedings of HM 2010 – Seventh International Workshop on Hybrid Metaheuristics, Springer, LNCS, vol 6373, pp 174–189

[228] Pirkwieser S, Raidl GR, Puchinger J (2007) Combining Lagrangian decomposition with an evolutionary algorithm for the knapsack constrained maximum spanning tree problem. In: Cotta C, van Hemert JI (eds) Evolutionary Computation in Combinatorial Optimization – EvoCOP 2007, Springer, LNCS, vol 4446, pp 176–187

[229] Pisinger D (1999) Core problems in knapsack algorithms. Operations Research 47:570–575

[230] Pitsoulis LS, Resende MGC (2002) Greedy Randomized Adaptive Search procedure. In: Pardalos PM, Resende MGC (eds) Handbook of Applied Optimization, Oxford University Press, pp 168–183

[231] Poojari CA, Beasley JE (2009) Improving Benders decomposition using a genetic algorithm. European Journal of Operational Research 199(1):89–97

[232] Pop PC (2002) The generalized minimum spanning tree problem. PhD thesis, University of Twente, The Netherlands

[233] Pop PC, Still G, Kern W (2005) An approximation algorithm for the generalized minimum spanning tree problem with bounded cluster size. In: Broersma H, Johnson M, Szeider S (eds) Algorithms and Complexity in Durham 2005, Proceedings of the first ACiD Workshop, King's College Publications, Texts in Algorithmics, vol 4, pp 115–121

[234] Potluri A, Singh A (2013) Hybrid metaheuristic algorithms for minimum weight dominating set. Applied Soft Computing 13(1):76–88

[235] Prandtstetter M, Raidl GR (2008) An integer linear programming approach and a hybrid variable neighborhood search for the car sequencing problem. European Journal of Operational Research 191(3):1004–1022

[236] Press WH, Teukolsky SA, Vetterling WT, Flannery BP (1992) Numerical Recipes in C: The Art of Scientific Computing, 2nd edn. Cambridge University Press

[237] Price KV, Storn RM, Lampinen JA (2005) Differential Evolution: A Practical Approach to Global Optimization. Springer

[238] Prim RC (1957) Shortest connection networks and some generalisations. Bell System Technical Journal 36:1389–1401

[239] Prüfer H (1918) Neuer Beweis eines Satzes über Permutationen. Archiv für Mathematik und Physik 27:742–744

[240] Puchinger J, Raidl GR (2007) Models and algorithms for three-stage two-dimensional bin packing. European Journal of Operational Research 183:1304–1327
[241] Puchinger J, Raidl GR, Pferschy U (2006) The core concept for the multidimensional knapsack problem. In: Gottlieb J, Raidl GR (eds) Evolutionary Computation in Combinatorial Optimization – EvoCOP 2006, Springer, LNCS, vol 3906, pp 195–208
[242] Puchinger J, Raidl GR, Pferschy U (2010) The multidimensional knapsack problem: Structure and algorithms. INFORMS Journal on Computing 22(2):250–265
[243] Raidl GR (2006) A unified view on hybrid metaheuristics. In: Almeida F, Blesa Aguilera MJ, Blum C, Moreno Vega JM, Pérez MP, Roli A, Sampels M (eds) Proceedings of HM 2006 – Third International Workshop on Hybrid Metaheuristics, Springer, LNCS, vol 4030, pp 1–12
[244] Raidl GR (2015) Decomposition based hybrid metaheuristics. European Journal of Operational Research 244(1):66–76
[245] Raidl GR, Hu B (2010) Enhancing genetic algorithms by a trie-based complete solution archive. In: Cowling P, Merz P (eds) Evolutionary Computation in Combinatorial Optimization – EvoCOP 2010, Springer, LNCS, vol 6022, pp 239–251
[246] Raidl GR, Julstrom BA (2003) Edge-sets: An effective evolutionary coding of spanning trees. IEEE Transactions on Evolutionary Computation 7(3):225–239
[247] Raidl GR, Puchinger J, Blum C (2010) Metaheuristic hybrids. In: Gendreau M, Potvin JY (eds) Handbook of Metaheuristics, International Series in Operations Research & Management Science, vol 146, 2nd edn, Springer, pp 469–496
[248] Raidl GR, Baumhauer T, Hu B (2014) Speeding up logic-based Benders' decomposition by a metaheuristic for a bi-level capacitated vehicle routing problem. In: Blesa MJ, Blum C, Voss S (eds) Proceedings of HM 2014 – Ninth International Workshop on Hybrid Metaheuristics, Springer, LNCS, vol 8457, pp 183–197
[249] Rechenberg I (1973) Evolutionsstrategie: Optimierung technischer Systeme nach Prinzipien der biologischen Evolution. Frommann-Holzboog
[250] Reeves CR (ed) (1993) Modern Heuristic Techniques for Combinatorial Problems. Wiley & Sons
[251] Reeves CR, Rowe JE (2002) Genetic Algorithms: Principles and Perspectives. A Guide to GA Theory. Kluwer Academic Publishers, Boston, USA
[252] Rei W, Cordeau JF, Gendreau M, Soriano P (2008) Accelerating Benders decomposition by local branching. INFORMS Journal on Computing 21(2):333–345
[253] Reich G, Widmayer P (1990) Beyond Steiner's problem: A VLSI oriented generalization. In: Nagl M (ed) Graph-Theoretic Concepts in Computer Science, LNCS, vol 411, Springer, pp 196–210

[254] Ribeiro Filho G, Nogueira Lorena LA (2000) Constructive genetic algorithm and column generation: an application to graph coloring. In: Chuen LP (ed) Proceedings of APORS 2000, the Fifth Conference of the Association of Asian-Pacific Operations Research Societies within IFORS

[255] Rios-Solis YA, Sourd F (2008) Exponential neighborhood search for a parallel machine scheduling problem. Computers & Operations Research 35(5):1697–1712

[256] Rodriguez FJ, García-Martínez C, Blum C, Lozano M (2012) An artificial bee colony algorithm for the unrelated parallel machines scheduling problem. In: Coello Coello CA, Cutello V, Deb K, Forrest S, Nicosia G, Pavone M (eds) Proceedings of PPSN XII – 12th International Conference on Parallel Problem Solving from Nature, Springer, LNCS, vol 7492, pp 143–152

[257] Ronald S (1994) Complex systems: Mechanism of adaptation. In: Stonier R, Yu XH (eds) Complex Systems: Mechanism of Adaptation, IOS Press, Amsterdam, pp 133–140

[258] Ropke S, Pisinger D (2006) An adaptive large neighborhood search heuristic for the pickup and delivery problem with time windows. Transportation Science 40(4):455–472

[259] Rossi F, van Beek P, Walsh T (2006) Handbook of Constraint Programming. Elsevier

[260] Rothberg E (2007) An evolutionary algorithm for polishing mixed integer programming solutions. INFORMS Journal on Computing 19(4):534–541

[261] Rubin SM, Reddy R (1977) The locus model of search and its use in image interpretation. In: Reddy R (ed) Proceedings of IJCAI 1977 – 5th International Joint Conference on Artificial Intelligence, William Kaufmann, vol 2, pp 590–595

[262] Ruiz R, Stützle T (2007) A simple and effective iterated greedy algorithm for the permutation flowshop scheduling problem. European Journal of Operational Research 177(3):2033–2049

[263] Ruthmair M, Raidl GR (2012) A memetic algorithm and a solution archive for the rooted delay-constrained minimum spanning tree problem. In: Moreno-Díaz R, Pichler F, Quesada-Arencibia A (eds) Proceedings of the 13th International Conference on Computer Aided Systems Theory: Part I, Springer, LNCS, vol 6927, pp 351–358

[264] Sabuncuoglu I, Bayiz M (1999) Job shop scheduling with beam search. European Journal of Operational Research 118(2):390–412

[265] Saraiva RD, Nepomuceno NV, Pinheiro PR (2013) The generate-and-solve framework revisited: Generating by simulated annealing. In: Middendorf M, Blum C (eds) Evolutionary Computation in Combinatorial Optimization, LNCS, vol 7832, Springer, pp 262–273

[266] Sha DY, Hsu CY (2008) A new particle swarm optimization for the open shop scheduling problem. Computers & Operations Research 35(10):3243–3261

[267] Shapira D, Storer JA (2002) Edit distance with move operations. In: Apostolico A, Takeda M (eds) Proceedings of CPM 2002 – 13th Annual Sym-

posium on Combinatorial Pattern Matching, LNCS, vol 2373, Springer, pp 85–98
[268] Shaw P (1998) Using constraint programming and local search methods to solve vehicle routing problems. In: Maher M, Puget JF (eds) Principle and Practice of Constraint Programming – CP98, Springer, LNCS, vol 1520, pp 417–431
[269] Shaw P, De Backer B, Furnon V (2002) Improved local search for CP toolkits. Annals of Operations Research 115:31–50
[270] Shen C, Li T (2010) Multi-document summarization via the minimum dominating set. In: Proceedings COLING 2010 – 23rd International Conference on Computational Linguistics, Association for Computational Linguistics, Stroudsburg, PA, pp 984–992
[271] Skiena SS (2008) The Algorithm Design Manual, 2nd edn. Springer
[272] Slaney J, Walsh T (2001) Backbones in optimization and approximation. In: IJCAI 2001 – Proceedings of the Seventeenth International Joint Conference on Artificial Intelligence, pp 254–259
[273] Solnon C (2010) Ant Colony Optimization and Constraint Programming. Wiley
[274] Spears WM, De Jong KA, Bäck T, Fogel DB, de Garis H (1993) An overview of evolutionary computation. In: Brazdil PB (ed) Proceedings of ECML 1993 – European Conference on Machine Learning, LNCS, vol 667, Springer, pp 442–459
[275] Stützle T (1999) Local Search Algorithms for Combinatorial Problems - Analysis, Algorithms and New Applications. DISKI – Dissertationen zur Künstlichen Intelligenz, Infix, Sankt Augustin, Germany
[276] Stützle T (2006) Iterated local search for the quadratic assignment problem. European Journal of Operational Research 174(3):1519–1539
[277] Stützle T, Hoos HH (2000) MAX-MIN ant system. Future Generation Computer Systems 16(8):889–914
[278] Subhadrabandhu D, Sarkar S, Anjum F (2004) Efficacy of misuse detection in ad hoc networks. In: Proceedings of IEEE SECON 2004 – First Annual IEEE Communications Society Conference on Sensor and Ad Hoc Communications and Networks, IEEE Press, pp 97–107
[279] Taillard ÉD (1991) Robust Taboo Search for the Quadratic Assignment Problem. Parallel Computing 17:443–455
[280] Talbi EG (2009) Metaheuristics: From Design to Implementation. Wiley & Sons
[281] Tang J, Guan J, Yu Y, Chen J (2014) Beam search combined with MAX-MIN ant systems and benchmarking data tests for weighted vehicle routing problem. IEEE Transactions on Automation Science and Engineering 11(4):1097–1109
[282] Thiruvady D, Meyer B, Ernst A (2011) Car sequencing with constraint-based ACO. In: Proceedings of GECCO 2011 – 13th Annual Conference on Genetic and Evolutionary Computation, ACM Press, pp 163–170

[283] Thiruvady D, Singh G, Ernst AT, Meyer B (2013) Constraint-based ACO for a shared resource constrained scheduling problem. International Journal of Production Economics 141(1):230–242

[284] Thiruvady D, Singh G, Ernst AT (2014) Hybrids of integer programming and ACO for resource constrained job scheduling. In: Blesa MJ, Blum C, Voss S (eds) Proceedings of HM 2014 – Ninth International Workshop on Hybrid Metaheuristics, Springer, LNCS, vol 8457, pp 130–144

[285] Thiruvady DR, Blum C, Meyer B, Ernst AT (2009) Hybridizing Beam-ACO with constraint programming for single machine job scheduling. In: Blesa MJ, Blum C, Di Gaspero L, Roli A, Sampels M, Schaerf A (eds) Proceedings of HM 2009 – Sixth International Workshop on Hybrid Metaheuristics, Springer, LNCS, vol 5818, pp 30–44

[286] Thomas E, Matúš M (2010) A (4+ ε)-approximation for the minimum-weight dominating set problem in unit disk graphs. In: Approximation and Online Algorithms, 8th Int. Workshop, WAOA 2010, Springer, LNCS, vol 6534, pp 135–146

[287] van Hoeve WJ, Hooker JN (eds) (2009) Proceedings of CPAIOR 2009 – 6th International Conference on the Integration of AI and OR Techniques in Constraint Programming for Combinatorial Optimization Problems, LNCS, vol 5547. Springer

[288] Vidal T, Maculan N, Ochi LS, Vaz Penna PH (2015) Large neighborhoods with implicit customer selection for vehicle routing problems with profits. Transportation Science DOI 10.1287/trsc.2015.0584

[289] Vimont Y, Boussier S, Vasquez M (2008) Reduced costs propagation in an efficient implicit enumeration for the 0–1 multidimensional knapsack problem. Journal of Combinatorial Optimization 15(2):165–178

[290] Vose MD (1999) The Simple Genetic Algorithm: Foundations and Theory. MIT Press

[291] Voudouris C (1997) Guided local search for combinatorial optimization problems. PhD thesis, Department of Computer Science, University of Essex, pp. 166

[292] Voudouris C, Tsang E (1999) Guided Local Search. European Journal of Operational Research 113(2):469–499

[293] Walshaw C (2002) A multilevel approach to the travelling salesman problem. Operations Research 50(5):862–877

[294] Walshaw C (2004) Multilevel refinement for combinatorial optimization problems. Annals of Operations Research 131:325–372

[295] Walshaw C (2008) Multilevel refinement for combinatorial optimisation: Boosting metaheuristic performance. In: Blum C, Blesa Aguilera MJ, Roli A, Sampels M (eds) Hybrid Metaheuristics – An Emerging Approach to Optimization, Studies in Computational Intelligence, vol 114, Springer, pp 261–289

[296] Walshaw C, Cross M (2000) Mesh partitioning: A multilevel balancing and refinement algorithm. SIAM Journal on Scientific Computing 22(1):63–80

[297] Wang F, Lim A (2007) A stochastic beam search for the berth allocation problem. Decision Support Systems 42(4):2186–2196

[298] Wang L, Wang SY, Xu Y (2012) An effective hybrid EDA-based algorithm for solving multidimensional knapsack problem. Expert Systems with Applications 39(5):5593–5599

[299] Wolsey LA (1998) Integer Programming. Wiley-Interscience

[300] Yagiura M, Ibaraki T (1996) The use of dynamic programming in genetic algorithms for permutation problems. European Journal of Operational Research 92(2):387–401

[301] Yuen SY, Chow CK (2007) A non-revisiting genetic algorithm. In: IEEE Congress on Evolutionary Computation, IEEE Press, pp 4583–4590

[302] Zakeri G, Philpott AB, Ryan DM (1999) Inexact cuts in Benders decomposition. SIAM Journal on Optimization 10(3):643–657

[303] Zaubzer S (2008) A complete archive genetic algorithm for the multidimensional knapsack problem. Master's thesis, Vienna University of Technology, Institute of Computer Graphics and Algorithms, Vienna, Austria

[304] Zha J, Yu JJ (2014) A hybrid ant colony algorithm for U-line balancing and rebalancing in just-in-time production environment. Journal of Manufacturing Systems 33(1):93–102

[305] Zhang W (2004) Configuration landscape analysis and backbone guided local search. Part I: satisfiability and maximum satisfiability. Artificial Intelligence 158(1):1–26

[306] Zou F, Wang Y, Xu X, Li X, Du H, Wan P, Wu W (2011) New approximations for minimum-weighted dominating sets and minimum-weighted connected dominating sets on unit disk graphs. Theoretical Computer Science 412(3):198–208